ENVIRONMENTAL HEALTH -
PHYSICAL, CHEMICAL AND BIOLOGICAL FACTORS

PERMAFROST

DISTRIBUTION, COMPOSITION AND IMPACTS ON INFRASTRUCTURE AND ECOSYSTEMS

ENVIRONMENTAL HEALTH - PHYSICAL, CHEMICAL AND BIOLOGICAL FACTORS

Additional books in this series can be found on Nova's website under the Series tab.

Additional e-books in this series can be found on Nova's website under the e-book tab.

ENVIRONMENTAL HEALTH -
PHYSICAL, CHEMICAL AND BIOLOGICAL FACTORS

PERMAFROST

DISTRIBUTION, COMPOSITION AND IMPACTS ON INFRASTRUCTURE AND ECOSYSTEMS

OLEG S. POKROVSKY
EDITOR

nova publishers
New York

For permission to use material from this book please contact us:
Telephone 631-231-7269; Fax 631-231-8175
Web Site: http://www.novapublishers.com

Additional color graphics may be available in the e-book version of this book.

Library of Congress Cataloging-in-Publication Data

Permafrost : distribution, composition and impacts on infrastructure and ecosystems / editors, Oleg S. Pokrovsky (University of Toulouse, France).

pages cm

Includes index.

ISBN 978-1-62948-830-1 (hardcover)

1. Permafrost. I. Pokrovsky, Oleg S.

GB641.P44 2013

551.3'84--dc23

2013047180

Published by Nova Science Publishers, Inc. † New York

Contents

PREFACE

Over the past decade, the scientific community's interest in the Arctic/Antarctic regions and permafrost has increased tremendously, producing articles that amount to the entire number published on these topics between 1950 and 2002 (Web of Science search). This is due to the tremendous importance of permafrost-dominated continental zones in the water and carbon cycle on Earth and the high vulnerability of permafrost regions to a warming climate. This book is intended to respond to the growing demands of academics and the general public for state-of–the-art research in permafrost science, in particular as it relates to impacts on infrastructure and ecosystems. Compared to more common non-permafrost aquatic and terrestrial environments, permafrost regions exhibit specific and unique features that render them among the most active physical, chemical and biological systems on the Earth's surface. One reason is their extremely short period of permafrost thaw during the year. During this period, multiple physical reorganizations occur in the soil and rock structure: freezing front migration cryoturbation, ice thaw and freeze. On the other hand, biotic components of permafrost systems are also most active during the several months of liquid water existence. These features produce extremely high fluxes of energy, water, and dissolved and suspended materials that occur within permafrost systems if normalized on a daily basis. A thorough and multidisciplinary approach is therefore prerequisite to an adequate and predictive understanding of these extremely complex systems.

Under climate change, permafrost regions are no doubt the most vulnerable and fragile. This is especially true for discontinuous/sporadic permafrost regions, where the majority of infrastructural and economic activity occurs. Due to a relatively thin frozen layer and ground temperatures close to zero degrees Celsius, these regions respond most quickly to a rise in air and ground temperature from a warming climate. Moreover, the effect of climate change on permafrost is strongly reciprocal; with the thawing of huge carbon stocks in frozen soils or methane hydrate releases, these greenhouse gas components will be released into the atmosphere, thus further accentuating climate change, not only at high latitudes but also across the entire planet. The detailed mechanisms of a possible "permafrost time bomb" are unknown at present, but knowledge and prediction of these mechanisms will require a detailed understanding of the elementary physical, chemical and biological processes that occur at both a local (mm to cm scale) and global (m to km) scale in permafrost-affected ecosystems. This book intends to bring together research from diverse but highly complementary scientific disciplines, to illuminate the main processes occurring in permafrost systems and to identify the possible mechanisms controlling fluxes of energy and matter at various scales.

The first chapter presents current knowledge about various cryogenic processes and their impacts on infrastructure. This information includes a general and well-illustrated description of permafrost's definition and abundance and the main features of its evolution, and it is written for a wide public audience. It contains all of the basic definitions and descriptions necessary to understand the subsequent, more specific chapters. Permafrost is perennially frozen ground that occurs naturally and has a temperature colder than 0°C (32°F) continuously for 2 or more years. Its thickness is related to the air temperature, soil characteristics, geothermal gradient, and geological history of an area. The main focus of the 1st chapter is to describe the cryogenic processes that take place in freezing and thawing ground and in permafrost under changing temperatures, as well as to describe the geological transitions that occur from melting ice. This chapter describes the distribution of permafrost on the Earth's surface and explains the problems that arise from the intersection of populations and permafrost; it also describes and illustrates the types of ground ice and explains the main cryogenic processes and their impacts on infrastructure, such as: frost heaving, thermokarst, thermoabrasion, thermoerosion, frost cracking, solifluction, and the formation of rock streams and rock glaciers. An important aspect of this chapter is that it shows how cryogenic processes affect human activities in northern regions. However, economic damage has not been very high due to the low quality development of infrastructure of territories under permafrost. The situation is likely to change due to the rise of industrial and infrastructural activity in Polar

Regions. As such, detailed studies of cryogenic processes are necessary to develop protection measures and to mitigate the effects of these processes.

The second chapter is devoted to the memory of Albert Lincoln Washburn (1911-2007), who contributed immensely to the fields of periglacial geomorphology and permafrost research. This chapter describes patterned ground, one of the most common morphologies formed in various periglacial environments, with or without permafrost. Patterned ground is usually formed by differential frost heave on superposed sediments with various frost susceptibilities. They may develop as self-organized nets, merging as coalescing forms, or as cracking guided nets (thermal or desiccation), defining contiguous forms that are perennially stable. Frost susceptibility develops in sediments and soils in the form of ice lensing that is controlled by the grain size, the freezing rate and the availability of moisture. Over time, frost shattering, stone jacking by frost and particle leaching at thaw may modify the frost susceptibility, limiting or enhancing the tendency for ice lensing. In terms of moisture availability, patterned ground primarily develops on shallow permafrost, which retains a perched water table in late summer, providing moisture for the winter frost. This chapter also presents a perspective on the evolution of pattern ground under a climate change scenario. With recent warming, the permafrost-table has mostly become deeper, starving the active layer of available moisture that drives a differential frost heave; therefore, the patterned forms of formerly drained areas inactivate; furthermore, those in a former waterlogged state come to the surface, deforming the layers. Lakes may drain, activating patterned ground on the freshly emerged bottom, but partly inactivating patterned ground fields located higher in the topography, such as in Svalbard. If precipitation also increases, the changes are limited. If cooling ensues, as in the 1970s, the active layer is shallower but waterlogged in late summer. Biological activity, especially in mountainous areas, may adapt to these change in drainage and frost heaves, leading to specific vegetation morphologies. In this case, patterned ground may be partly biogenic as in temperate mountains.

The third chapter continues the discussion of patterned ground in mountainous regions by presenting a zone of optimum development of patterned ground; this zone is characterized by semi-arid regions and indicated by permafrost-lines, and references sub-glacial permafrost below the historic arid cold-based Tibetan Ice Sheet. It synthesizes the results collected since the 1970s during field research in the Arctic (Spitsbergen (Svalbard), Greenland and Alaska) as well as in Tibet and the high mountains of the North- and South-hemispheres. The global uniformity of permafrost and periglacial indicators are proven by their existence not only in high arctic latitudes but also in subtropical mountains. Pingos, as well as polygonal frost cracks, have been met between 28 and 35°N above c. 4600 - 5300 m asl. They are considered to be the key forms of arctic periglacial and permafrost areas. The periglacial form of patterned ground can be found everywhere there is constantly thawed soil, namely where mountain summits do not reach the permafrost limit anymore. In systematic contrast to the Arctic, where the lower border of permafrost is not even reached from above, numerous mountains in temperate, subtropical and tropical latitudes not only reach the lower limit of permafrost from below but also tower above this limit by up to several 1000 m. In the semi-arid to arid subtropical high mountains, the size of the individual forms of patterned ground (i.e., their diameter and the depth of their sorting of material) not only decreases in the ever-thawing-soil with diminishing absolute altitude this applies to all extra-arctic mountains – but also decreases in areas with permafrost soil that has increasing absolute altitude going higher The investigations described in this chapter show that important parts of the nearly 2 million km^2 of extended ground moraines on the glacial Tibetan ice sheet were formed under subglacial permafrost conditions. This supports past research of cold-based inland ice. Overall, 7 types of combinations of periglacial levels with patterned ground and glacial levels on the globe can be differentiated.

The 4th chapter is devoted to state-of-the-art hydrogeochemical (numerical) modeling of heat and water transfers in permafrost-dominated watersheds in Central Siberia, which are heavily affected by continuous permafrost coverage. This chapter addresses the thermo-hydrological determinants of weather fluxes in boreal watersheds and offers insights into what may be expected from a mechanistic modeling approach. The main foci of this chapter are the weathering fluxes in the permafrost-dominated watersheds of the basaltic region of the Putorana Plateau in Central Siberia, one of the largest and most pristine boreal forest areas. First, a qualitative model of thermo-hydrological function of watershed surfaces in this region is presented. A discussion about the couplings of water, energy and matter fluxes emphasizes the importance of phase changes and advective transfers. In the second part of this chapter, after a short review of published numerical tools dealing with the coupled transfer of water and energy in the active layer of permafrost areas, perspectives associated with the use of modern numerical techniques are discussed. A modeling effort is introduced, which has been initiated in OpenFOAM®, an open source toolbox for computational fluid dynamics that allows multi-physic coupling and massive parallel computing.

The results illustrate the importance of snow cover and the moss/soil organic horizon thickness on the thermo-hydrological behavior of the active soil layer depth of this area.

The second half of the book comprises 2 chapters devoted to numerical descriptions of permafrost thaw lakes using remote-sensing techniques a biogeochemical study of thermokarst lakes in western Siberia and a chapter on permafrost biota, focusing on the fungi. The 5th chapter addresses the remote study of thermokarst lake dynamics in west-Siberian permafrost. It presents the features driving the thermokarst processes in frozen bog-lake systems under today's global warming trend. Thermokarst lakes are considered to be a suitable indicator of geocryological changes in permafrost landscapes using remote-sensing.

Methodological issues for remote measurement of lake areas based on optical and radar satellite images are discussed and the intra-seasonal dynamics of thermokarst lake areas were studied using radar images. The long-term dynamics of thermokarst, based on satellite images from Landsat, were carried out at test sites located in different permafrost areas and in different landscape zones and revealed the growth trends of the total area of thermokarst lakes in West-Siberian permafrost from a south to north direction. An increase was observed in the total area of lakes in the northern part of continuous permafrost above 70 degrees N. In the southern part of continuous permafrost (south of the 70th latitude), a reduction tendency of the total area of thermokarst lakes area was observed. This can be interpreted as a displacement of the current border between continuous and discontinuous permafrost to the north under the influence of global warming.

The 6th chapter presents modeling and the prediction of thermokarst lake field dynamics using satellite images. It is based on an analysis of thermokarst lake fields in permafrost areas in Western Siberia, which was carried out using the remotely sensed measurement of lake areas. On the basis of experimentally determined statistical properties of the thermokarst lake fields, the geo-simulation model of the dynamics of these fields was proposed. The relationship between climatic and geocryological parameter changes was studied and a multiple regression equation was developed to take into account the temperature and precipitation changes in the model. An application of the model allowed for forecasting the dynamics of thermokarst lake fields until 2030 and 2050 with a linear trend in the mean annual temperature of air, given by the Intergovernmental Panel on Climate Change (IPCC). The experimental model showed that the lake areas in West-Siberian permafrost will be reduced, on average, if the mean annual air temperature continues to increase in the coming decades.

The 7th chapter addresses the impact of permafrost thaw on the biogeochemistry of thermokarst lakes in western Siberia, combined with a multidisciplinary approach for assessing the current status of lakes and predicting possible future changes.

In this chapter, the biogeochemical and morphological features of thermokarst lakes were analyzed based on their natural evolution: from frozen palsa depressions to large, km-size lakes, to drained lakes. Based on a compilation of ~120 analyses of these water bodies, the average concentration of organic carbon and the major and trace elements in Western Siberia's thermokarst lakes were reported, and the development of their chemical composition, exchange with the atmosphere and their effect on the riverine fluxes that run from the land to the ocean under climate warming were predicted. The accelerating permafrost thaw and rising water temperatures in this region will likely shorten the life cycle of the thermokarst thaw, increase the fluxes of both CO_2 and CH_4 into the atmosphere, increase the concentration and delivery of dissolved organic carbon and related trace metals into the hydrological network and increase the potential bioavailability of micronutrients. As a result, destabilization of discontinuous/sporadic permafrost regions will be ongoing and the retroactive link between climate change and thermokarst lake biogeochemistry may be more significant than is currently expected.

Finally, the last chapter diverges from the rest of the book by presenting a detailed analysis of *Penicillium* fungi from permafrost, the biosynthesis of secondary metabolites and the peculiarities of its growth and development. It is based on a study of secondary metabolite biosynthesis with a strain isolated from permafrost and from its usual habitat. Interestingly, no principal differences between the two strains were found, but the strains isolated from permafrost exhibited lower growth rates. This chapter contains an extensive, up-to-date, compilation of secondary metabolite profiles and taxonomic positions of fungi strains extracted from permafrost environments.

Taken together, the 8 chapters of this book provide a comprehensive, up-to-date description and analysis of the basic geomorphological, physical, hydrological, chemical and biological aspects of permafrost-affected ecosystems, their interaction with other components of the landscape and their impact on human life and infrastructure.

Possible future changes will include the destabilization of frozen ground, an increase in the number of small thaw lakes and a decrease in the surface area of large lakes, the enhancement of the greenhouse gas emissions and inevitable changes in soil and aquatic biota. A concerted, multidisciplinary approach that brings together scientists from various disciplines with different spatial and temporal scales of investigation is necessary to foresee the impact of permafrost thaw on human society and the Earth.

In: Permafrost
Editor: Oleg S. Pokrovsky
ISBN: 978-1-62948-830-1

Chapter 1

CRYOGENIC PROCESSES AND THEIR IMPACT ON INFRASTRUCTURES

S. M. Govorushko
Pacific Geographical Institute, Vladivostok, Russia
Far Eastern Federal University, Vladivostok, Russia

ABSTRACT

Information on distribution of permafrost on the globe is given. Common problems of interactions between humanity and permafrost are considered. Different types of ground ice (porous, segregated, wedge, buried, sublimation, and pingo ice) are described. Main cryogenic processes (frost heaving, thermokarst, thermoabrasion, thermoerosion, frost cracking, solifluction, rock streams, and rock glaciers) are discussed. The mechanisms of these processes; their impacts on human activities (industrial and civil site development; water, air, railway, automobile, and pipeline transport; mineral resource industry; hydropower engineering; agriculture, and so on); mitigation measures; and other topics are considered.

INTRODUCTION

Permafrost is perennially frozen ground, a naturally occurring material with a temperature colder than 0°C (32°F) continuously for 2 or more years.

Its thickness is related to the air temperature, soil characteristics, geothermal gradient, and the geological history of an area. Cryogenic processes are those that take place in freezing and thawing grounds, and in permafrost grounds under conditions of changing temperatures and the rocks' transitions through the melting of ice.

1. Distribution of Permafrost

The area of the cryolithozone (permafrost zone) of the Earth is 38.15 million km^2, which corresponds to 25.6% of the land surface; 21.35 million km^2 fall in the northern hemisphere. Permafrost underlies 25.6% of Earth's land area, including probably all of Antarctica, 99% of Greenland, 80% of Alaska, 61.5% of Russia, 55% of Canada, and 20% of China (Govorushko 2012). Permafrost is completely absent in the continent of Australia, while, in Africa, it is possible only in the high mountain areas. A substantial part of the present-day permafrost was inherited from the last glacial period, and now it is slowly thawing.

The maximum thickness of the permanently frozen ground is 1,493 m (4,898 ft) in the northern Lena and Yana River basins in Siberia (http://en.wikipedia.org/wiki/Permafrost). In North America, the observed thickness of frozen rocks in northern Alaska reaches 740 m (http://www.britannica.com/EBchecked/topic/452187/permafrost/65728/Climatic-change).

At lower latitudes, permafrost exists at high elevations; it is referred to as Alpine permafrost. The largest area of Alpine permafrost, 1.5 million km^2 (580,000 sq mi), exists in western China. Alpine permafrost in the contiguous United States is present on about 100,000 km^2 in mountainous areas of the west. It occurs at elevations as low as 2,500 m in the northern states and at about 3,500 m in Arizona (http://www.britannica.com/EBchecked/topic/452187/permafrost/65728/Climatic-change). In South America in the Andes along the Atacama Desert, permafrost begins at an altitude of 4,400 m (14,400 ft) and is continuous above 5,600 m (18,400 ft) (http://en.wikipedia.org/wiki/Permafrost).

Large areas of permafrost also lie under the Arctic Ocean, and on the northern continental shelves of North America and Eurasia. These areas are referred to as subsea or offshore permafrost. Seasonally frozen rocks are more widely distributed. They occupy vast territories except in regions with tropical and subtropical climates. The distribution of permafrost on the globe is shown in Figure 1.

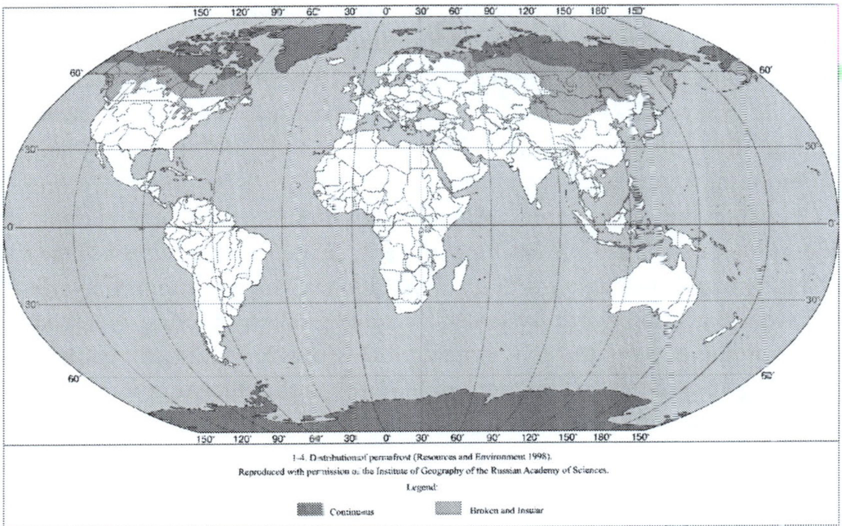

Figure 1. Map of permafrost.

2. Problems of Interactions between Humanity and Permafrost

The presence of permanently frozen grounds in itself complicates human life and activities in cold regions. Few or no kinds of business activity take place in these regions; that is, all kinds of construction, agriculture, geological surveys and geological exploration, extraction of commercial minerals, operation of different means of transport, military activities, and other activities are severely hindered by the frozen ground.

Interest in economic development of areas with permafrost led to the emergence of the science of **geocryology,** which later was subdivided into the sciences of **general geocryology** (theory and study of general laws of development of permafrost), **engineering geocryology** (study of behavior and properties of permafrost as they relate to economic activity), and **agrobiological geocryology** (study of frozen rocks and soils as they relate to agriculture and forestry).

Construction on permafrost is difficult because the heat of the structure (buildings, pipelines, etc.) can thaw the permafrost and destabilize the structure. Two alternative **principles of construction** in permafrost regions are (1) with preservation of land in the frozen state; and (2) without conservation of permafrost. The second approach has three variants (Lomtadze 1977): (1) gradual thawing in the course of construction and maintenance of facilities; (2) artificial thawing prior to erection of the structures; and (3) substitution of frozen ground with thawed ground.

Choosing the construction principle is determined by engineering and economic expediency. *Conservation* of land in the frozen state is generally attained by using cooling units that are divided into surface (underfloor spaces, ventilation ducts, ductwork systems, etc.) and sunken (cooling pipes, shafts, etc.). Communication lines that generate heat (pipelines, electric cables, etc.) can be placed in the open air (Figure 2). With gradual thawing, it is necessary to reduce irregularities in settlement by (1) supporting uniform thawing (thermal insulation, special heaters, etc.); and (2) preventing abrupt changes in loading. For construction in areas with perennially frozen ground, foundations in the form of posts, piers, or end-bearing piles are usually used (Lomtadze 1977).

The occurrence of frozen rocks, on its own, complicates production from placers considerably. Dredgers developing deposits in areas with perennially frozen rocks are characterized by low key performance indicators. It is necessary to pre-thaw frozen rocks to the depth of the metallic stratum or to locate taliks (unfrozen pockets). Losses of metal due to incomplete extraction of frozen stratum sometimes reach 50–70%. Incomplete provision of dredgers with thawed resources occurs especially during the first half of flushing season.

Frost-bound rocks are characterized by high strength, which hampers extraction from open-pit and underground mines. On the other hand, due to permafrost grouting the rocks, kimberlite pipes have been used effectively in Yakutia in quarries with nearly vertical walls (Figure 3). One more advantage of permafrost is the possibility of long-term storage of products.

Reservoirs of gas hydrates can form in permafrost; these are accumulations of gas (often methane) bound with water at the molecular level. During the formation of these compounds at low temperatures and increased pressures, methane molecules are transformed into crystals of hydrates, forming solid matter that is similar in consistency to loose ice. As a result of molecular compacting, one cubic meter of natural methane hydrates in solid form contains about 164 m^3 of methane in the gaseous phase and 0.87 m^3 of water (Vinogradova 2001). Accumulations of gas hydrates occur in the cryolithic zone and in the near-bottom part of ocean sediments, predominantly along the eastern and western margins of the Pacific Ocean and in the eastern margins of the Atlantic Ocean, at depths of 300–400 to 1,000–1,200 m (Figure 4).

Photograph credit: http://en.wikipedia.org/wiki/Permafrost, 29 June 2007.

Figure 2. Pipes in the permafrost cannot be dug into the ground as they are in warmer climates, so they are raised over the ground and are insulated.

Photograph credit: Stepanov Alexander, 17 July 2004.

Figure 3. The open pit of the Udachnaya Diamond Mine, Russia, from a helicopter.

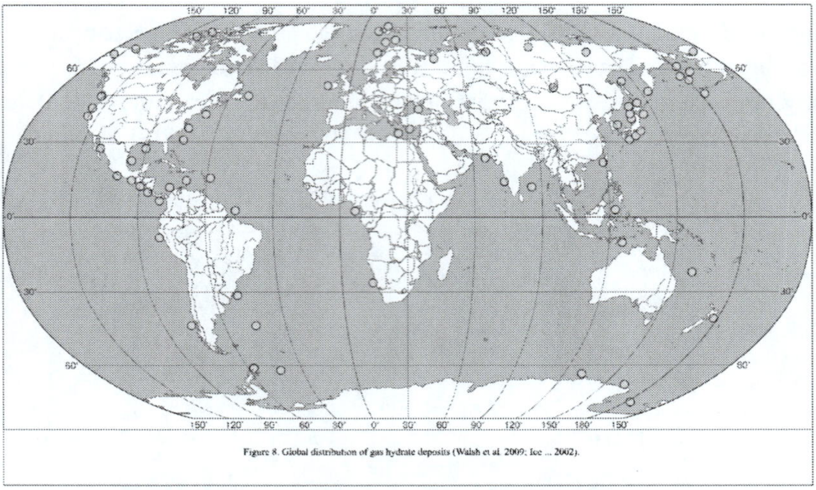

Figure 4. Gas hydrates map.

3. Types of Ground Ice

More than 20 classifications of underground ices have been developed. These classifications can be found in books of Vtyurin (1975), Vtyurina and Vtyurin (1970), and Yershov (2002). The classifications are based on the differences in the mechanisms of ice formation, dimensions, shape, occurrence, and structure. Attention is focused on them because the ice content of permafrost is the most important feature of permafrost affecting human life in the north. There are five main types of ice in perennially frozen ground: (1) porous ice; (2) segregated, or Taber, ice; (3) foliated, or wedge, ice; (4) pingo, or bulgunniakh, ice; and (5) buried ice.

1. Porous ice fills (full or partially) pore spaces in the ground. It is formed by pore water freezing in situ with no addition of water (Figure 5). The ground contains no more water in the solid state than it could hold in the liquid state.
2. Segregated, or Taber, ice consists of almost pure ice that often exists as an extensive horizontal layer. It can range in thickness from hairline to more than 15 cm. Segregated ice commonly occurs in alternating layers of ice and soil (Figure 6). Pore ice and segregated ice occur both in seasonally frozen ground and in permafrost.

3. Wedge ice, or foliated ice (Figure 7) is vertically oriented ground ice that extends into the top of a permafrost layer. These features are approximately 2 to 3 m wide at their top and extend into the soil about 8 to 10 m. It forms in cracks that develop in the soil during winter because of thermal contraction. In the spring, these cracks fill with liquid water from melting snow, which subsequently re-freezes. The freezing process causes the water to expand in volume, increasing the size and depth of the crack. The now-large crack fills with more liquid water, and again it freezes, causing the crack to enlarge further. This process continues for many cycles until the ice wedge reaches its maximum size.
4. Pingo, or bulgunniakh, ice is clear, or relatively clear (Figure 8). It occurs in permafrost more or less horizontally or in lens-shaped masses, and it originates from groundwater under pressure.
5. Buried ice is ice formed on the surface (sea, lake, river, or glacier ice) and buried under mineral or organic-mineral sediments.

Photograph credit: S. P. Davydov (North-East Scientific Station, Pacific Geographical Institute, Cherskiy, Russia), May 2005.

Figure 5. Poreous ice is ice which fills or partially fills pore spaces in permafrost; forms by freezing soil water in place, with no addition of water. Photo shows pore ice in the core (downstream of Kolyma River).

In addition, sublimation ice is formed by reverse sublimation of water vapor onto cold surfaces (Figure 9). It is relatively insignificant, however.

The amount of ice in permafrost throughout the world is estimated at between 200,000 and 500,000 km^3 (49,000 to 122,000 cu mi). An estimated 10% by volume of the upper 3 m of permafrost on the northern Coastal Plain of Alaska is composed of foliated ground ice (ice wedges). Taber ice, the most extensive type of ground ice, comprises 75% of the ground by volume in some areas. The pore and Taber ice content at depths of 0.5–3 m (the top 0.5 m is seasonally thawed) is 61% by volume, and the content at depths of 3–9 m is 41%. Pingo ice accounts for less than 0.1% of the permafrost. In the Arctic Coastal Plain of Alaska, the total ice content in the permafrost is estimated at 1,500 km^3, and at depths greater than 9 m most of it is present as pore ice (http://www.britannica.com/EBchecked/topic/452187/permafrost/65728/Climatic-change).

The composition of the ice in different areas varies greatly. For example, our calculations of volumetric ice content of frozen rocks in the eastern Yana-Indigirka plain (Russia) gave a value of 58%, including 40% wedge ice and 18% segregated and **pore** ice or ice-cement (Govorushko 1981). The ice content in the frozen rocks can be as much as 90%.

Photograph credit: S.P. Davydov (North-East Scientific Station, Pacific Geographical Institute, Cherskiy, Russia), June 2013.

Figure 6. A segregated ice consists of almost pure ice that often exists as an extensive horizontal layer. The ice layer grows because of the active migration of water from around the feature. Segregated ice in the core from depth 185-200 cm (downstream of Kolyma River) is shown here.

Photograph credit: S.M. Govorushko (Pacific Geographical Institute, Vladivostok, Russia), July 1975.

Figure 7. Wedge ice or foliated ice is the term for large masses of ice growing in thermal contraction cracks in permafrost. Photo shows ice wedges on downstream of Indigirka River (Russia). Distance between the axes of vein strikes is 8-12 m, width in the roof is 1-2 m while, at the depth of 13-15 m, the veins extend to 4-6 m. In the top left part of photo, the researcher taking the soil samples is seen.

Photograph credit: I. V. Dorogoy, Institute of Biological Problems of the North, Magadan, Russia.

Figure 8. Pingo or bulgunniakh ice is clear. It occurs in permafrost more or less horizontally or in lens-shaped masses and originates from groundwater under pressure. Photo shows pingo ice in upper reaches of Levy Yarakvaam River (Chukotka, Russia).

Photograph credit: S.P. Davydov (North-East Scientific Station, Pacific Geographical Institute, Cherskiy, Russia), July 2007.

Figure 9. Sublimated ice is ice originating from gas moisture without passing through an intermediate liquid phase. Photo shows sublimated ice (downstream of Kolyma River).

4. Main Cryogenic Processes and Their Impacts on Infrastructure

The number of cryogenic processes is quite high; those that have the greatest impact on human activities include frost heave, thermokarst, thermal abrasion, thermal erosion, cryogenic cracking, solifluction, rock streams, and rock glaciers.

Those processes with the greatest relief-forming effects and volumes of reworked material include (Geoecology of the North 1992) (1) thermokarst; (2) thermal erosion; and (3) thermal abrasion. Thermokarst can reach 6.5 x 10^6 m^3/km^2; thermal erosion can remove up to 0.3 x 10^6 m^3/km^2; thermal abrasion, up to 7 x 10^3 m^3/km^2; fast solifluction, up to 4 x 10^3 m^3/km^2; and solifluction, from 30 to 400 m^3/km^2. In the plains of North Russia, the denudation intensity (i.e., average drops in the earth's surface level) reaches 7 mm/yr (Voskresensky et al. 1999). The basic cryogenic processes are considered in the following sections.

4.1. Frost Heaving

Frost heaving is upward movement of the soil surface caused by an increase in its volume during freezing, due to the spreading of particles by growing ice crystals. The **intensity** of the swelling depends on the degree of water saturation, and it is especially high when the moisture content increases through inflow from neighboring areas.

Cryogenic heaving is most characteristic of the clayey silts of the Quaternary age. In some areas, this process occurs on a large scale. For example, in the Ob-Nadym interfluves, 85–90% of the territory is affected by frost heaving (Abaturova et al., 2009).

The heaving **intensity** depends on the degree of water saturation; the intensity is particularly high in "open systems," where the moisture content increases due to exogenous inflow. It is evident as humps of 2–4 to 50 cm high and up to 10–12 m in diameter.

4.1.1. Mechanisms of frost heaving

In determining the **mechanism** of the influence of frost heaving on engineering facilities, the tangential and normal forces of a swelling are identified. When freezing occurs near the foundation, the ground freezes to its side face. The swelling forces tend to move the foundation up, together with a layer of frozen ground. If the forces of ground freezing with the foundation are less than the mass of the structure, then the frozen layer moves relative to the foundation. The shear strength of the foundation when it freezes along with the ground determines the *tangential forces* of swelling.

When a frozen layer increases in thickness, the force of ground freezing with the foundation can exceed the load resistance. In this case, there will be "bulging" of the foundation; that is, its heave together with the ground will result in loss of stability and normal operation of the structure. The normal swelling forces act at right angles to the foundation. The straight freezing of the swelling ground near the side faces of a foundation results in their all-round pressurization. When this occurs, a swelling nonuniformity can lead to one-sided pressure and horizontal displacement. The soil freezing under a foundation determines the development of *normal forces of swelling* at its foot.

Under the action of the forces of frost ground heaving on a foundation secondary stresses arise in the bearing members of the structure and result in *deformations*; these deformations can disturb the normal operation of the building or make it unusable. Deformations can cause the formation of cracks in foundations, ceilings, floors, and walls, and skewing of door and window openings. These deformations have a cyclic, seasonal nature and repeat every year (Figure 10). During the spring melting of swelling ground, water permeability and compressibility increase, while the carrying capacity decreases, which results in differential settlement of a building.

4.1.2. Impacts on engineering structures

Frost heaving represents a danger for **motor roads** and **railroads** (Figure 11) and for **airfields,** causing disruptions in their continuity and evenness. These disruptions, in turn, can lead to emergency conditions in transport due to pushes and strokes in the course of its motion (bursting of rails, automobile accidents, and aircraft accidents on takeoff, etc.).

A survey of the central section of the Baikal-Amur Mainline 5–6 years after construction was completed showed detrimental deformations caused by heaving along 11% of the railroad. In addition, frost heaves of relatively small amplitudes were found over 55% of the length of the railroad (Geocryological Dangers 2000).

In Norway, 300 km of railroads go out of service due to frost swelling every year. In the United States, the railroads in the states of Wisconsin, North Dakota, Nebraska, and Idaho are affected, to the maximum extent, by this phenomenon (Larionov 1974).

Frost heaving causes serious problems for residential and industrial construction. It causes damage especially in areas of deep seasonal freezing; for example, in Transbaikalia and North Kazakhstan. In 1955–1956, many one-story houses were constructed on strip foundations with a laying depth of 70 cm in three railway stations in North Kazakhstan. About 1–2 yr later, about 90% of the houses showed deformations coming through rents in the walls up to 10 mm wide (Orlov et al. 1987). In this book, many instances of building deformation due to frost heaving are listed: residential buildings in the cities of Khabarovsk and Yuzhno-Sakhalinsk and at the station in Taiga (Tomsk Oblast); a district hospital building in the city of Sretensk; a barn in the village of Mankovo; and a motor repair shop in the state farm of Kopunsky (all in the Chita region).

Frost heaving also constitutes a certain danger for **communication and transmission lines, bridges,** and other structures. In addition, the phenomenon of heaving and the resulting collapse of piles, posts, etc., is widely known. For example, at the Skovorodino cryogenic station of the Central Scientific Research Institute of Construction, a post was pushed up by 220 cm in 38 years (Kotlov 1978). The intensity of the heaving is illustrated in Figure 12. The center of one of the bridges in the Alaskan Railroad rose by

35.5 cm during the winter of 1952–1953. In order to replace the rails in their original position, the upper piles had to be cut (Anderson and Trigg 1981).

Photograph credit: S.M. Govorushko (Pacific Geographical Institute, Vladivostok, Russia), March 2013.

Figure 10. The photo shows the deformed fence in the Vladivostok suburb. Because a footing of poles was less than the frost zone thickness, their heaving happened. The uniformity of heaving caused by differences in the ground water saturation and composition resulted in the fence deformation.

Photograph credit: V.S. Afanasenko, Department of Geocryology, Moscow State University, Russia

Figure 11. Mounds of heaving do not have considerable impact on human, since the lands of their presence are usually sparsely inhabited and poorly developed. The photo shows mounds of heaving on the Tynda-Zeysk section of the Baikal-Amur Railroad, Russia.

Swelling is a primary cause of underground **pipeline** deformation, especially where the pipes cross rivers (Figure 13, 14). For example, in November 1972 through January 1973, a pipe break at a weld accompanied by a gas release happened as a result of frost swelling in a section of the Messoyakha-Norilsk pipeline where it crossed the Yenisey River (Atlas of Natural and Technogeneous Dangers and Risks 2005).

The hazard is also high where pipelines cross waterlogged areas. For example, the annual values of ground swelling along the Urengoy-Nadym gas pipeline are 3 to 147 mm. Overriding occurred in a gas pipeline section with a diameter of 1,220 mm and a wall thickness of 20 mm by 16–86 mm. On the pipe, additional bends of 29–86 mm long appeared along a bent part of the pipe measuring 36–60 m. In some Yakut gas pipelines, vertical movements of pipes reach 229 mm (Geocryological Dangers 2000). In the Yamburg gas field (Northern Siberia, Russia), more than 3,000 piles supporting pipelines are repaired every year because of frost heave damage (Titkov and Ablyazina 2008).

Frost heaving also has adverse effects on **grassland farming** and **crop production**. During freezing, the soil (especially loose soil) is slightly raised, and as a result, the roots of plants are detached. After melting, the soil subsides and plants with detached roots remain under the sun and wither.

To some extent, frost swelling also adversely affects **hydropower engineering**. The straight freezing of clayey dam cores results at times in destruction of their watertight integrity (Natural-Anthropogenic Processes and Environmental Risk 2004).

In regions where permafrost is present, perennial mounds caused by cryogenic heaving (pingos or bulgunniakhs) are abundant. These are circular to elongate ice-cored mounds that form by injection and freezing of pressurized water in near-surface permafrost. There are two types of pingos, based on origin. The closed-system type forms in level areas when unfrozen groundwater in a thawed zone becomes confined on all sides by permafrost, freezes, and heaves the frozen overburden to form a mound. This type is larger and occurs mainly in tundra areas of continuous permafrost. The open-system type is generally smaller and forms on slopes when water beneath or within the permafrost penetrates the permafrost under hydrostatic pressure. A hydrolaccolith (water mound) forms and freezes, heaving the overlying frozen and unfrozen ground to produce a mound. The process of pingo formation lasts from several tens to several hundreds of years (Figure 15, 16).

Photograph credit: F.M. Rivkin, OJSC «Fundamentproekt», 27 August 2007.

Figure 12. The intensity of heaving depends on the water saturation of grounds. The heaving of the casing pipes of the observation holes in the experimental site (1968-1973) of the Institute "Fundamentproyekt" in Labytnangi, Tyumen Region, Russia is shown.

Photograph credit: V.P. Sheshenya, OJSC «Fundamentproekt», 1974.

Figure 13. Frost heaving (a rising of the soil surface caused by an increase in its volume in the course of freezing, due to the spreading of particles by growing ice crystals) is a serious problem for pipelines. This picture shows Messoyakha-Norilsk gas pipeline crossing a small river. There are pills with cross-bars, on which are placed the pipes.

Photograph credit: F.M. Rivkin, OJSC «Fundamentproekt», 1984.

Figure 14. Ten years later due to frost heaving total buckling of pills was 1.6-1.8 m. When height of pills increased cross-bars were cut and welded to a new level. Later adjustable cross-bars were installed but they gradually also were overaged. The same place is shown.

Photograph credit: S.M. Govorushko (Pacific Geographical Institute, Vladivostok, Russia), August 1976.

Figure 15. Pingos are hills of soil-covered ice pushed up by hydrostatic pressure in an area of permafrost. They grow in areas of abundant water supply. First stage of origin of pingo in Northern Yakutia (Russia) is shown here.

Photograph credit: Emma Pike.

Figure 16. The largest pingos mount more than 50 m high and 600 m in diameter. The photo shows the pingos near Tuktoyaktuk, Northwest Territories, Canada.

There are more than 11,000 pingos on Earth (Grosse and Jones 2011). Their distribution within the cryolithic zone is very nonuniform. For instance, within a 40,000 km^2 area on the western Arctic Coastal Plain of northern Alaska, 1,247 pingo forms were identified, ranging in height from 2 to 21 m, with a mean height of 4.6 m. Pingos in this region are of hydrostatic origin, with 98% located within 995 drained lake basins, most of which are underlain by thick aeolian sand deposits. Morphometric analyses indicate that most pingos are small to medium in size (<200 m diameter), gently to moderately sloping (<30 deg), circular to slightly elongate, and of relatively low height (2 to 5 m). However, 57 pingos stand higher than 10 m, 26 have a maximum slope greater than 30 deg, and 42 are larger than 200 m in diameter (Jones et al. 2012). The *Kadleroshilik Pingo* (or *Kadleroshilik Mound*) is the highest known pingo in the world. It located about 40 km (25 mi) southeast of Prudhoe Bay in the U.S. state of Alaska. It measures 54 m (178 ft) high (Mackay 1998). Large pingos also can be found around the Mackenzie River estuary. They reach 50 m high and 600 m in diameter.

Since frost heaving is observed in less-developed regions of the world, damage related to it for the present is not great.

4.1.3. Protection measures

The following kinds of **anti-heaving measures** have been identified (Orlov et al. 1977): (1) reclamation engineering; (2) constructional; (3) physico-chemical; and (4) combined. *Engineering amelioration measures* are subdivided into thermal amelioration and hydrotechnical amelioration. *Thermal amelioration* is aimed at increasing the temperature of frozen ground and decreasing the frost depth, which reduces tangential forces and weakens their intensity. Toward that end, thermal insulation is installed at foundations, or communications cables are buried near the foundations, which generate heat in the ground.

Hydro land reclaiming measures are focused on lowering groundwater levels and decreasing the water content of the ground. Collecting ditches, chutes, trenches, etc., are used to drain foundation soils during the summer and autumn seasons.

Constructional measures are focused on improving a foundation's operating efficiency. Their goal is to decrease the frost heaving force and to adapt the foundations and aboveground parts of the structures to nonuniform deformations. To reduce tangential forces of frost heaving, pier and pile foundations are used instead of strip and heavy foundations; the foundations are extended downward and covered with grease, and their side faces are smoothed. In order to adapt the structures to nonuniform deformations, the structures are reinforced (reinforced-concrete booms are placed in the walls, foundations are framed; and other procedures are used). *Physical-chemical measures* consist of treating the ground with binding materials to make it watertight and cause it to lose its heaving properties. Saturating the ground with salt solutions is another technique; this process lowers the freezing temperature and, accordingly, decreases the frost depth.

Combined methods consist of various combinations of the techniques discussed above.

The effects of cryogenic heaving on human activities are considered in Table 1.

Table 1. Effects of cryogenic heaving on infrastructure

Basic objects	Nature of the effect	Consequences of the effect	Mitigation measures
Railroad transport	Deformation of the land surface (frost heaves) due to nonuniform migration of moisture during freezing	Distortion of transversal and longitudinal profiles of the roadbed, shocks to trains when they are in motion, damaging the wheels, and rupture of the rails	Grading of the roadbed, drainage, replacement of heaving grounds, speed restriction, provision of temporary routes, etc.
Highway and air transport	Deformation of the land surface (frost heaves) due to nonuniform migration of moisture during freezing	Breach of continuity and smoothness of airstrips and highways, bumps, vibration, reduction in load-bearing capacity, fast depreciation of transport	Arrangement of watertight interlayers and heat-insulating blankets, electro-osmotic drying of the roadbed, replacement of heaving grounds, etc.
Industrial and civil development, telephone lines, bridges	Deformation of the land surface (frost heaves) due to nonuniform migration of moisture during freezing	Deformation of structures due to nonuniform heaving and following subsidence, collapse of walls and floorings, heaving of foundations and posts	Making foundations deeper than the freezing depth, extension of foundations downwards, their covering with greases, gravel packings etc.
Plant cultivation, animal raising	Deformation of the land surface (bulgunniakhs, humps-burial grounds, hydrolaccoliths) due to nonuniform migration of moisture to the freezing front	Deterioration of land cultivation conditions (food and fodder crops)	Reduction in depth of seasonal frost penetration, lowering of the head of intrapermafrost and infrapermafrost waters
Plant cultivation, animal raising	Stretching effect on plant roots due to rise of soil surface in the course of freezing	Drying of crops due to breakage of plant roots	Well-timed rolling of seeds subjected to heaving to encourage the growth of new roots
Hydraulic power industry	Increase in volume of wet grounds due to freezing	Disturbance of watertight stability of loamy dam core	Heat-insulation materials, warming, salinization of core material

4.2. Thermokarst

The term **thermokarst processes** means a melting of ground ice accompanied by strain in beds (initiation of subsidence and depressions or formation of cavities in these beds). Thermokarst processes occur extensively in the depositional plains of northern Eurasia and North America. Countries with the most **extensive development** of thermokarst include Russia, Canada, and the United States.

4.2.1. Causes of thermokarst

Causes of thermokarst can be classified as general and particular (Popov et al. 1985). *General causes* include (1) climate warming; (2) intensification of climate continentality that increases the depth of summer thawing; and (3) other physiographic factors, such as afforestation and increases in the depth of snow cover. *Particular causes* include (1) cracks (frost, dynamic); (2) trampling of vegetation cover by people and animals; (3) forest fires; (4) forest fell; (5) construction (roads, structures); and (6) plowing.

The development of thermokarst can be subdivided into three stages (Foundations of Geocryology 2001, Vol. 4). During the *first stage,* owing to changes in external heat exchange, the depth of seasonal thawing increases and the surface depression fills up with water. In the *second stage***,** the depth of the reservoir begins to exceed the critical value at which the average annual temperature of the bottom surface becomes 0°C. The *third stage* comes when the diameter of the lake exceeds the permafrost thickness and a talik forms under it.

This scheme characterizes the development of lake thermokarst. Later on, the lake may drain owing to inburst of water to a nearby river valley. Extensive, frequently treeless flat-bottomed hollows called alases (Figure 17) are typical of many northern regions.

In the case of thermokarst accompanied by water runoff, thermodenudation microrelief arises, such as bajdzherahi, or cemetery mounds (Figure 18), and thermocirques. Depending on the volume of melted ice, the sizes of negative landforms are very different and depths of craters can reach several tens of meters (Figure 19).

Photograph credit: S.M. Govorushko (Pacific Geographical Institute, Vladivostok, Russia), June 1975.

Figure 17. Alas is a shallow depression which is formed by subsidence of the permafrost due to repeated melting and refreezing. An alas first develops as a shallow lake as melt water fills the depression. The lake eventually dries out and is replaced by grasses and other herbaceous vegetation. Alases are often used for pasturage for horses as well as hay-fields. Alas in lower reaches of Indigirka River (Russia) is shown here.

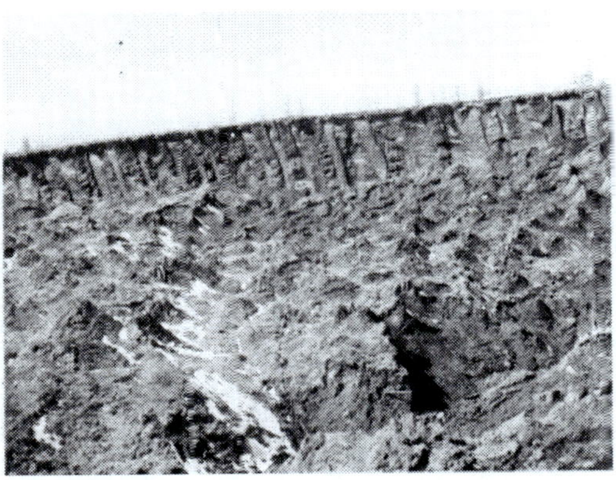

Photograph credit: S.M. Govorushko, July 1979.

Figure 18. Bajdzherahi are hillocks with heights of up to 5 meters and diameters of up to several tens of meters often arranged in zigzag order. They form from grounds of the central parts of sites as a result of ice-veins melt-out in the course of thermokarst. The Bajdzherahi in lower reaches of Indigirka River (Russia) are shown.

Photograph credit: United States Geological Survey.

Figure 19. The extra large thermokarst funnel in Alaska, USA (near the Fairweather Glacier) is shown.

4.2.2. Intensity of Thermokarst

In different areas of the cryolithic zone, thermokarst intensities differ considerably. These differences are caused by a number of factors. For example, low compartmentalization of territory and, consistently, poor drainage, prevents the evacuation of sagging material and thermokarst develops further. The dependence of thermokarst intensity on surface compartmentalization is demonstrated by comparison of the southern Yamal Peninsula with the the Taz Peninsula. These areas are located in the same natural zones (forest-tundra, typical and south tundra), but volumes of reworked material in the Taz Peninsula are 2–2.5 times those in the Yamal Peninsula due to greater compartmentalization of the surface in the Taz Peninsula. Under such conditions, the thermokarst hollow transforms into erosion thermokarst and begins to actively develop. Continuous removal of material contributes to the further thawing of rocks and increases in its depth.

For northern Russia as a whole, the following laws of thermokarst development have been identified (Voskresensky et al. 1999). Minimal volumes of reworked material are found in the high north (Arctic tundra) and south (northern taiga and forest-tundra) of the cryolithic zone. Relatively large but infrequent thermokarst formations are characteristic of the Arctic tundra. In the northern taiga and forest-tundra, numerous small thermokarst formations arise, which, on the whole, also lead to an insignificant amount of subsidence. Reductions in area and increases in the number of hollows southwards are also characteristic of other areas of the cryolithic zone of Russia (e.g., the Primorsk plain and the plains of Chukotka).

Quantitative characteristics of thermokarst are presented in Table 2.

4.2.3. Impacts on engineering structures

Thermokarst constitutes a serious danger to the safety, stability, and normal operation of **railroads** and **motor roads** (Vtyurin and Govorushko 2012). The total length of railways laid on permanently frozen soils in Russia is about 5,000 km, and about 15% of the track undergoes permanent deformation (Figure 20), which leads to restrictions in train speeds.

Table 2. Quantitative characteristics of thermokarst in West Siberia*

Region	Quantity, forms/km^2	Size, km^2	Depth, m	Total volume, million m^3/km^2
Northern part				
Yamal Peninsula	0.3–1	0.07–0.25	6–12	0.5–1.[illegible]
Taz Peninsula,	8–10	0.25–0.32	2–4	1–2
Central part				
Yamal Peninsula, southern part	0.6–5	0.31–0.32	11–15	2–3
Salekhard plain	1–5	0.33–0.36	13–15	3–[illegible]
Southern part	4–12	0.027–0.25	3–6	0.5–1.[illegible]

* Voskresensky et al. (1999).

Photograph credit: N.F. Grigor'ev (Institute of Permafrost Studies, Yakutsk, Russia), 1967.

Figure 20.Thermokarst poses a formidable threat to railroad maintenance. The photo illustrates the numerous deformations of the Salekhard-Nadym railroad in the section between the switching track of Rastushchy and station of Poluy. The railroad was constructed in the 1930s, however, the construction was not completed.

Photo credit: OJSC «Fundamentproekt», 1997.

Figure 21. Thermokarst subsidence deforms underground pipelines. Photo shows a disruption of anchor clips due to “break surface” pipeline on Urengoy gas field. The pipeline was originally laid in a shallow trench to it through the anchors were attached concrete blocks. To the output of the pipeline on the surface has a combination of Archimedes bioyant force and pressure of cryogenic groundwater within the halo of thawing. Grass and soil fragments on the surface of the pipe is evidence that earlier pipeline was buried, the rest of the soil fell under the tube.

Photo credit: Y.A. Murzin, Institute of Permafrost Studies, Russian Academy of Sciences, July 1993.

Figure 22. The photo shows the wall collapse in the dwelling house in Yakutsk, Russia caused by thermokarst. Nobody was injured. To prevent such collapses, houses are to be built on piles. Thereby, the air space under the house excludes the heat impact on the frozen ground. This house was erected ‘low-sitting’, and, for the long time of exploitation, the air space got stuffed with finely dispersed material. This led to gradual melting of frozen grounds lying below.

Construction in 2001–2006 of the Qinghai–Tibet railway (Qinghai–Xizang) in China is the newest stage of railway construction in the cryolithic zone. To construct the railway, engineering solution were sought that allowed stability of the track on icy, collapsing (during defrosting) permafrost. Fifty percent of the 1,142-km-long railroad passes through permanently frozen grounds having average annual temperatures of –0.5° to –3.6°C and thicknesses of 5–25 to 60–130 m and more. Observations made after the railroad was put into operation showed a continuous increase in thermokarst subsidence of the roadbed. The total lengths of track with ground subsidence were 15.76 km in 2005 and 18.56 km in 2007 (Kondratyev 2012).

Thermokarst sometimes contributes to serious accidents. For example, in the summer of 1984, subsidence of the Tynda-Berkakit village railroad body base near the village of Magot, Russia, took place due to thawing of ice-saturated ground. As a result, the rail track was destroyed and a train was derailed (Atlas of Natural and Technogeneous Dangers and Risks 2005).

An inspection of the asphalt Norilsk-Talnakh motor road (constructed in the 1960s through 1970s), and a section of railroad from the Valyok quay to the Golikovo station (constructed in 1935), demonstrated that deformations of the automobile and rail road beds developed over not less than 50% of their lengths. Deformations were characterized by different forms and scales. Over every 3 km, one section with deformations endangered the safety of traffic (Isakov 2012).

Thermokarst subsidence also deforms surface and underground **pipelines** (Figure 21), frequently resulting in accidents. Thermokarst also causes great problems for **residential** (Figure 22) and **industrial engineering**. According to A. I. Dementyev's data (Kotlov 1978), 64% of all deformations in buildings in areas of permanently frozen ground are caused by thermokarst alone. For example, substantial deformations occurred in nine-story residential buildings constructed in 1978 through the 1980ss in Norilsk (Lolayev 1998). Many instances of building deformations in the cities of Magadan, Vorkuta, and Norilsk are also given by V. D. Lomtadze (1977). Practically all the buildings erected in Magadan oblast (Russia) prior to 1951 (when they were constructed without regard for the frozen subsoil properties) were deformed due to ground bearing capacity failure as a result of thawing (Russian Arctic 1996).

The cases of deformations of foundations due to thermokarst in the city of Anadyr in the 1980s are widely known. As a result, many auxiliary structures of the local thermal power plant (TPP) were destroyed, and its basic structures were endangered (Myagkov 1995). The cause of damage to the buildings was generally the formation of a thawing basin, resulting in irregular settlement of foundations and, as a consequence, initiation of cracks, subsidence of quoins, warping of door frames, etc.

The presence of thermokarst results in increases in reservoir volumes over those that are projected (according to different estimates, up to 15% and more). These increases occur due to bottom subsidence and growth in dead storage, which delays the attainment of normal headwater levels, complicates reservoir operation conditions, and reduces electric energy generation. Defrosting of perennially frozen rocks near the Ust-Khantaika Hydroelectric Power Plant lasted 19 years. Thermal subsidence was also observed at the Vilyuy hydroelectric complex: during the first 4 years, perennially frozen rocks thawed under the dam, affecting them to a depth of 9 m; after 4 years, the effects were felt to depths of 6–9 m and the greatest depth of thawing reached 14 m (Malik 2005).

Thermokarst development on territories adjacent to lakes and reservoirs degrades water quality. Soluble materials released from degrading permafrost are transported to lakes and reservoirs by surface runoff, elevating concentrations of those materials in the water. Studies carried out in small upland catchments (<20 ha) between Inuvik and Richards Island, Northwest Territories, Canada, showed that mean concentrations of calcium, magnesium, and sulfate in lakes with watersheds affected by thermokarst are 8–20 times higher as compared with those in areas where this process does not develop (Kokelj et al. 2005).

4.2.4. Positive significance of thermokarst

In a number of cases, thermokarst has **beneficial effects** and is suitable for the following uses (Tomirdiaro 1978): (1) *hay lands* and *pastures* (emptying of lakes creates hollows with high meadow grass stand); (2) *road construction* (road building along the bottoms of thermokarst lakes is possible with much lower embankment heights because the underground ices have been destroyed by thermokarst and ground has been well compacted in the talik under the lake); (3) *civil construction* (construction of buildings in alases does not require preservation of the permafrost in the foundation, which reduces expenditures related to installation of communications); (4) high-quality *construction grounds* (grounds of perennially frozen rocks require long drying and compaction, while taliks under a lake are depleted of excessive moisture and compacted); (5) *water supplies* (water reservoirs-digouts in taliks under lakes are cheaper and more reliable than pressure water reservoirs, as neither thermoabrasion migration over the plain around dams nor breakthroughs through the system of the ice veins are a threat).

Table 3. Effects of thermokarst on infrastructure

Basic objects	Nature of the effects	Consequences of the effects	Mitigation measures
Industrial and civil development	Abrupt and nonuniform ground subsidence	Cracks in foundations, subsidence of corners of buildings, skewed door frames, etc.	Heat-insulating banking, ventilated under-floor spaces, replacement of grounds of foundations, etc.
Motor, railway, and pipeline transport	Abrupt and nonuniform ground subsidence	Deformation of beds of roads and railways, damage to pipelines	Rock dumping, ventilation tubing, channels, water removal
Hydraulic power industry	Abrupt and nonuniform ground subsidence	Cracks in the bodies of dams with possibility of their subsequent failure	Rock dumping on slopes, use of refrigerating units, liquid nitrogen, etc.
Animal raising	Outcrop of thawed deposits in the bottoms of emptied thermokarst lakes	Development in the hollows of high meadow grass stands suitable for use as hay fields and pastures	Blockage of drainage channels after 25–30 years and repeated turning into lakes for 2–3 years to prevent the substitution of valuable plants with cotton grass and mosses.
Automobile roads, civil engineering, underground utility systems	Outcrop of thawed deposits in the bottoms of emptied thermokarst lakes	Improvement of conditions and reduction of the costs of construction in connection with lack of underground ices, and presence of well-compacted ground in the under-lake talik	
Automobile roads, civil engineering, underground utility systems	Outcrop of thawed deposits in the bottoms of emptied thermokarst lakes	Possibility of production of high-quality construction grounds free of excessive moisture and with well-compacted ground	
Water supply facilities	Thermal effects of water resulting in formation of talik under the thermokarst lake	Possibility of carrying out dredging operations for development of water reservoirs-digouts.	Creation of earth bank around digout in order to decrease the inflow of lake waters enriched in humic substances and installation of gateways for passage of clean water during rains and snow melting

The most radical way to **prevent deformations** of structures is permanent *preservation of the frozen state of grounds* under them. *Other measures* can be used, such as the strengthening of grounds, filling of forming cavities with cement and sand, and pre-construction land development.

Effects of thermokarst on human activities are outlined in Table 3.

4.3. Thermoabrasion

The term **thermoabrasion** means a process of destruction of shores composed of perennially frozen rocks or ice due to the heating effects of water. Thermoabrasion (thermal abrasion) is an important process in forming the shores of Arctic seas (primarily in Russia, the United States, and Canada). Distribution of thermoabrasive shores is shown in Figure 23.

The length of the Russian segment of the Arctic ocean coast is 39,440 km (Stolbovoi 2002); that is, thermoabrasion affects 40% of the continental coastline of the Arctic seas of Russia (Romanovsky 1993). Thermoabrasion is of great significance in the development of thermokarst lakes and water reservoirs in zones of perennially frozen ground.

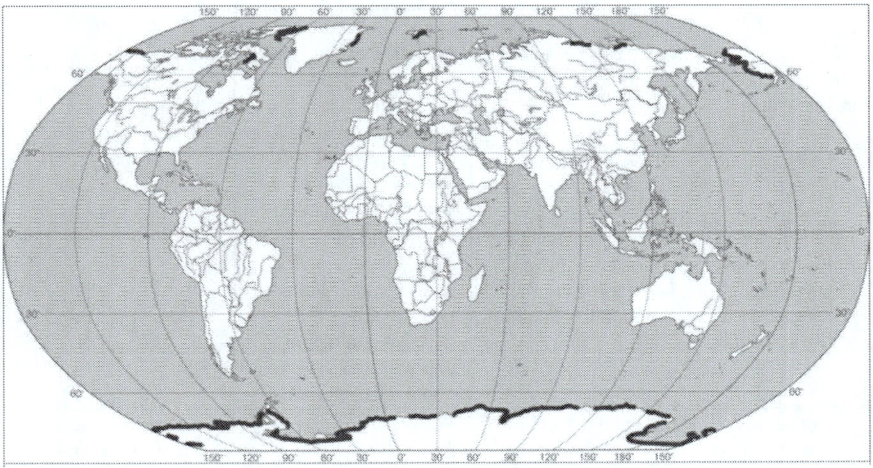

Figure 23. Map of thermoabrasion shores.

4.3.1. Mechanisms of thermoabrasion

The **basic process** of thermoabrasion is a washout of the underwater shoreface under the action of roughness and currents. It results in the formation of a niche, and further deepening causes frozen rock blocks to fall. The rate of thermoabrasion depends on the lithologic composition (the likelihood of washing out of rocks increases in the following order: clays, loams, clay sands, sands) and the ice content in the rocks (the greater the ice content, the higher the erosion rates). Even at subzero temperatures, the hard rocks are exposed to only mechanical abrasion.

4.3.2. Intensity of thermoabrasion

The **intensity** of thermoabrasion is determined by the ratio of thermal and mechanical energies to the rock erosion index. The *heat energy* of waves is proportional to the temperature of the surface water layer, while mechanical energy is proportional to the square of the wave heights and their speeds (Foundations of Geocryology 2001, Vol. 4). All other things being equal, the erosion of rocks increases in the following sequence: clays, clay loams, sandy clays, sands. Information on real values of thermoabrasion in different regions is presented in Table 4.

The total value of thermoabrasion for the Russian segment of the Arctic is estimated at 338 million t/yr; this much sediment comes to the coastal zone owing to thermoabrasion (Stolbovoi 2002). The volume of deposits entering the Laptev Sea due to washout of the islands in the Lena River delta reaches 1.8 million t/yr (Grigor'ev and Schneider 2002). A number of in situ observations have been aimed at estimating the losses of land. So, according to data of J. Brown and J. Jorgenson (2002), an 11 km sector of the shore near Barrow (northwestern Alaska) lost 28.2 ha during a period of 50 years.

The intensity of thermoabrasion is extremely high within an 85 km section of Anabaro-Olenek coast of the Laptev Sea, where about 170 km^2 of land was washed away over a period of 22 years (Are 1985).

The number of observations for *lake thermoabrasion rates* is much lower. According to S. V. Tomirdiaro (1978), the long-term average annual velocities of retreat of lake shores in the Anadyr tundra are 2–6 m/yr. Recession rates of icy shores of lakes in Central Yakutia reach 7–10 m/yr.

Table 4. Rates of thermoabrasion in different areas

Area	Rate of thermoabrasion, m/yr	Period	Source
Northwestern delta of Lena River (Laptev Sea)	0.2–1.5; on average, 0.6	1970–2001	Grigoryev et al. (2002)
Barrow-Kaktovik (northwestern Alaska)	0.5	1950–2001	Brown and Jorgenson (2002)
West Yamal	2	—	Kamalov et al. (2002)
West Yamal	0.3–3.2; on average, 0.4	—	Vasilyev (2002)
West Yamal	>1.0	—	Yuryev (2009)
Kharasavei Cape (Yamal Peninsula)	5.0	1953–1982	Geoecology of the North (1992)
Lena River delta	4.7	—	Grigor'ev and Schneider (2002)
Gulf of Anadyr	2–10.5	—	Lyubomirov (1996)
Kara Sea coast	2–3	—	Are (1985)
Laptev Sea coast	4–6	—	
East Siberian Sea coast	5–7	—	
Beaufort Sea coast	7–8	—	
Kara Sea coast (near settlement of Kara geological exploration)	3.0	—	Natural-anthropogenic processes and environmental risk (2004)
Kara Sea coast (Nyavotalova River mouth)	1.5	—	
Varandey Island (Barents Sea)	3–4	1987–1999	

The intensity of land elimination on reservoirs may also be extremely high. For example, over 25 years of the Bratsk Reservoir (Russia) storage operation, thermoabrasion has destroyed about 270 km^2 of the coast (Theoretical Basics of Engineering Geology 1985). Here, events of extremely high intensity were recorded. So, in 1962–1967, the shore retreated by 759 m near the Artumei settlement, and the erosion rates reached 435 m/yr and 150 m/d (Myagkov 1995).

4.3.2. Impacts on engineering structures

Thermoabrasion affects the following **kinds** of human activity: (1) industrial and civil site development; (2) water transport; (3) pipeline transport; (4) mineral resource industry; (5) hydropower engineering; and (6) agriculture.

The effects on **site development** are expressed as a threat to beach installations. In September 1986, a sharp intensification of thermoabrasion on the Alaskan coast of the Chukchi Sea took place as a consequence of two storms. The boroughs of Barrow and Wainwright experienced serious losses (Figure 24). In the first settlement, 152 people were evacuated and, later, 32 houses were transported to a new site (Walker 2001). Several power transmission line poles also had to be moved and, in addition, the storm damaged an archaeological monument: peat houses (Walker 1991).

Thermoabrasion threatened to destroy the Kharsavey polar station and a similarly named lighthouse in the Yamal Peninsula. Constructed in 1953, they were demounted and reestablished after a lapse of 30 years (Geoecology of the North 1992). Shore erosion in the Varandey industrial area (Nenets Autonomous Okrug) threatens the existence of the settlement, petroleum base, and airport (Natural-Anthropogenic Processes and Environmental Risk 2004).

Effects on **water transport** involve changes in navigation conditions. Thermoabrasion processes result in a reduction in depths and create problems for shipping. Water transport is also affected by the demolition of lighthouses and navigation markers. In addition, thermoabrasion causes problems where **underwater pipelines** make landfall. The influences of thermoabrasion on the **mineral resource industry** are rather positive and lie in the fact that, to a large degree, it forms offshore placer deposits of minerals.

Photo credit: H.J. Walker, July 1987.

Figure 24. The average rate of thermoabrasion does not exceed 0.5-1.0 m/year; however it may mount as high as 10 m/year. Coast retreat mostly occurs during 2-3 summer months; the process dramatically intensifies at times of heavy storms. The photo shows the coast of the Chukchi Sea in Alaska, nearby Wainwright Settlement. The severe storm of October 1986 exposed ice wedges and by that speeded up coastal destruction which imperiled dwellings.

The impact on **hydropower engineering** lies in the fact that thermoabrasion creates an abundance of solid particles. This causes the sedimentation of reservoirs and reduces their usable storage. When woody and peaty shores are destroyed, there is also clogging of waterways and chemical pollution. The effects on **agriculture** are expressed as the destruction of croplands and grazing lands; however, considering the small scales of this kind of human activity in the regions subjected to thermoabrasion, the effects are considered to be minor.

Thermoabrasion makes a considerable contribution to **global warming**. In the course of shore destruction, enormous amounts of carbon, methane, and other gases that had been sequestered in the frozen ground are released.

A complex multivariate analysis is used for the **forecasting** of thermoabrasion; the analysis requires determination of shore age and tendencies of ocean level changes, examination of the continental slope profile, determination of basic hydrodynamic factors, and knowledge of sea ice conditions and the effects of thermal factors (temperatures of water and air, solar radiation) on the stability of coastal perennially frozen rocks. The forecasting of thermoabrasion is based on the solution of problems for three interrelated processes: thawing of rocks, their subsidence, and erosion (Gevorkyan 2012).

The influence of thermoabrasion on human activities is outlined in Table 5.

4.4. Thermoerosion

Thermoerosion is a process that causes the breakup of frozen rocks. Simultaneous thermal and mechanical actions of water flows result in intrusion of the water flow into the frozen mass, with the formation of furrows, ruts, and cavities.

4.4.1. Mechanisms of thermoerosion

Thermoerosion is initiated where the vegetation cover is discontinuous, which can be caused by both natural factors (frost crack formation, solifluction, slip-outs, etc.) and anthropogenic factors. For thermoerosion to develop, the following conditions are necessary (Foundations of Geocryology 2001, Vol. 4): (1) presence of perennially frozen ground; (2) a grade of more than 1.5 deg; and (3) sufficient rainfall intensity. Critical precipitation values for the emergence of thermoerosion are different for different regions. For example, the active development of primary washouts in the Chara basin (Russia) is observed when rainfall is greater than 20 mm/d (Poznanin 1995). Thermoerosion occurs with greatest intensity where slopes are greater than 4.5 deg and daily rainfall exceeds the monthly mean.

The **intensity**of gully erosion is high. Elongation of gullies occurs at rates of 1–2 to 5–7 m/yr, reaching, in some cases, 20–30 m/yr, while, within ravines and hollows, they can be up to 100–150 m/yr.

Table 5. Effects of thermoabrasion on infrastructure

Basic objects	Nature of the effect	Consequences of the effect	Mitigation measures
Industrial and civil development, water transport	Thermal effect of water on the ice-containing rocks	Decay of buildings, lighthouses, and navigation marks due to collapse of coastal blocks and solifluction intensification	Accounting of process dynamics in the design, ramping of slopes, wave-cut walls etc.
Sea transport	Thermal effect of water on the ice-containing rocks	Complication of navigation due to change of depths in the coastal zone	Ramping of slopes, wave-cut walls, etc.
Pipeline transport	Thermal effect of water on the ice-containing rocks	Complication of landfall of submerged pipelines	Accounting for thermoabrasion dynamics in the design
Mining industry	Thermal effect of water on the ice-containing rocks	Formation of coastal placers of mineral resources	
Hydraulic power industry	Thermal effect of water on the ice-containing rocks	Decrease in electric energy generation due to changes in the usable storage during silting	Ramping of slopes
Water supply	Thermal effect of water on the ice-containing rocks	Chemical contamination of water during destruction of woody and peaty coasts	
Crop raising, grassland farming	Thermal effect of water on the ice-containing rocks	Destruction of arable and grazing lands	

The **mechanisms** of thermoerosion vary with the type of rock. *Sandy frozen rocks* are eroded by way of separation and migration of individual particles, while *clay rocks* are eroded when microaggregates and aggregates are washed away; the materials are decomposed into mineral particles and lose adhesion with each other.

Turfs tend to not be susceptible to erosion, as they possess specific structural adhesion properties owing to the abundance of plant roots. And thermoerosion is practically absent on the peaty soils of northern Europe because of this as well (Fundations of Geocryology 2001, Vol. 4).

Thermoerosion occurs due to the joint effects of thermal and hydromechanical factors on the degraded grounds. Until water flow has a thermal impact, frozen ground retains its resistance to erosion as compared with similar non-frozen rocks.

Thermoerosion is unlikely to occur when the eroding flow is at 0°C or when heat-insulating—for example, peaty—ground that reduces the thermal effect on the underlying permafrost is exposed. However, when thermal effects develop, the erosion intensity increases owing to the following reasons (Geoecology of the North 1992): (1) water flow increases due to ice melting; (2) the flow profile changes due to ground subsidence during thawing; and (3) the ground loosens in the course of thawing.

4.4.2. Kinds of thermoerosion and their intensity

Thermoerosion is subdivided into two **types**: bed and gully. The mechanism of *bed thermoerosion*is, to a large extent, similar to that of thermoabrasion. When a shore is being undercut, thermoerosion niches are formed (Figure 25), followed by the fall of blocks (Figure 26). When *gully thermoerosion* (Figure 27) develops, gravitational failures result in blockages in channels and, as a consequence, intense cutting and detachment of sides.

Data on the intensity of channel thermoerosion in different areas are given in Table 6.

One example of intense development of ravine thermoerosion owing to anthropogenic impacts is provided by E. Z. Kuchukov and E. D. Yershov (Foundations of Geocryology 2001, Vol. 4). During construction in the Salemal settlement, located on the Ob River terrace, moss and vegetation cover was eliminated within a large area due to the use of tractors and other construction machinery. Over three summer seasons, eight ravines measuring 100–250 m long, and numerous short ones with a total length of more than 1.2 km, appeared there.

The **intensity** of ravine thermoerosion is very important. Elongation of ravines is from 1–2 to 5–7 m/yr, though in some cases it reaches 20–30 m/yr; it can reach 100-150 m/yr in some cases (Sukhodrovsky 1979). In the early 1960s, the rate of ravine formation owing to thermoerosion in the territory of Salekhard and a number of neighboring villages reached 130 m/yr over 3 years; the rates decreased to 5–25 m/yr only in the following years (Russian Arctic 1996).

Detailed examinations of ravine thermoerosion in northern West Siberia were carried out by K. S. Voskresensky and coauthors (1999). This region includes a large expanse of land from north to south (about 700 km), and shows diverse landscape conditions, at least some of which are similar to those in other northern regions.

As to the volumes of material reworked by thermoerosion (in million cubic meters per square kilometer), they identified 10 areas in northern West Siberia and divided these areas into three groups based on the extent of thermoerosion reworking of the surface.

Thermoerosion to a weak extent is characterized values from 0.01 to 0.1 million m^3/km^2. These areas are situated within the forest-tundra, south tundra, and, partially, typical tundra. The maximum density of ravine formation here reaches s6/km^2, the average length is 100 to 400 m, and the depths of incision are 2–6 m.

The medium extent is characterized by values from 0.1 to 1.0 million m^3/km^2. Such volumes are common for the south tundra of Yamal and the typical tundra of the central Gydan Peninsula. These areas are essentially divided into parts by ravines; the number of ravines per unit area increases to 6–10, their average length is 400–600 m, while the depths of incision increase to 10–12 m.

For thermoerosion to a strong extent, values of 1.0 to 2–3 million m^3/km^2 are characteristic. Such areas are concentrated in the Arctic tundra and typical tundra in northern Yamal as well as in the central and northern Gydan Peninsula.

The densities of ravines are from 10/km^2 to 12/km^2 in some sectors. Their average length is 800–1,600 m, while depths of incision are 15–18 m. It should be noted that the density of ravines 12/km^2, is maximal for northern West Siberia. At this density, their heads practically reach the watersheds and the surface acquires the "badland" image.

Photograph credit: S.M. Govorushko, July 1975.

Figure 25. The nature of riverbed thermoerosion is in many ways similar to that one of thermoabrasion. Coastal cut-down forms thermoerosion niches. Such niche is shown in lower reaches of Indigirka River (Russia).

Photo credit: H.J. Walker, July, 1966.

Figure 26. Afterwards, the block movement takes place under action of gravity. The photo demonstrates the process of coast destruction in the Kolville River, Alaska, the USA.

Photo credit: A.N. Kozlov, Department of Geocryology, Moscow State University, Russia.

Figure 27. Thermoerosion also intensifies in case of human-related breach of vegetation cover. The construction of a pipeline and parallel power line triggered thermoerosion processes along the pipeline which threatened the balance of power transmission towers.

Table 6. Rate of retreat of coasts due to channel thermoerosion

Area	Value	Period	Source
Lena River (from Yakutsk to Aldan)	On the average, 6.5 m/yr	—	Are (1985)
Lena, Indigirka, Yana Rivers	20–30 m/yr	—	Are (1985)
Colville River delta (Alaska)	1–12, on the average, 1.6 m/yr	1949–1986	Walker et al. (1987)

Therefore, thermoerosion in northern West Siberia is distinguished by a well-defined zonal dependence. This is reflected in the sizes of the ravines and is definitely characterized by the volumes of reworked material, which reach maximum values in the Arctic tundra (more than 2 million m^3/km^2), while volumes in the forest-tundra and north taiga show minimum values (less than 0.01 million m^3/km^2) (Voskresensky et al. 1999).

4.4.3. Impacts on engineering structures

Riverbed thermoerosion affects different installations located within the coastal zone (harbor installations, transmission and communication lines, roads, pipelines, and other structures). H. J. Walker (2001) uses as an example the thermoerosion effects on the Nigilik village in the Colville River delta (Alaska, U.S.). From 1949 to 1986, a shore retreated there by more than 50 m and a house was threatened. In order to prevent the destruction of the house, it was transported over a distance of 30 m from shore.

Thermoerosion can cause significant displacements of pipelines. For example, the level of pipeline in some spots in the Medvezhye gas field (North Siberia, Russia) dropped by more than 3 m over a period of 34 years due to thermoerosion (Yegurtsov et al. 2011).

To mitigate the effects of thermoerosion, conserving frozen ground or delaying the defrosting of frozen ground is necessary. As for channel erosion, different structures are also used to help prevent the shore base from washing away, and to redirect the erosive flow to the opposite shore.

Effects of thermoerosion on human activities are outlined in Table 7.

4.5. Cryogenic (Frost) Cracking

Cryogenic (frost) cracking is a dissection of a frozen rock mass with cracks that develop when temperatures fall. It occurs in regions of both permafrost and seasonally frozen rocks.

4.5.1. Mechanisms of cryogenic cracking

Two types of cryogenic cracking have been identified (Grechishchev et al. 1980): (1) in the course of frost penetration; and (2) after the ground is frozen. The cracks form during the fall through winter period. They are most pronounced in areas with an acutely continental climate and insignificant snow depths. The **main cause** of frost cracking is strains connected with changes in the volumes of frozen soils; these changes are caused by temperature gradients and changes in the state of water in the rock massif. The *widths and depths of cracks* depend on the composition of the rocks, their uniformity, and temperature distribution. Their maximum lengths reach tens and hundreds of meters, while depths are 5–6 m. The widths of cracks at the top are generally 2–4 cm, though cracks more than 10 cm wide occur (Glaciological Encyclopaedia 1984).

Table 7. Impacts of thermoerosion on infrastructure

Basic objects	Nature of the effect	Consequences of the effect	Mitigation measures
Industrial and civil development, water transport	Thermal effect of water on ice-containing rocks	Decay of buildings, lighthouses, and navigating marks due to downfall of coastal blocks and solifluction intensification	Accounting of process dynamics in the design, ramping of slopes, wave-cut walls, etc.
River transport	Thermal effect of water on ice-containing rocks	Complication of navigation due to changes in depths within coastal zone	Ramping of slopes, wave-cut walls, etc.
Pipeline transport	Thermal effect of water on ice-containing rocks	Complication of onshore landfall of submerged pipelines	Accounting for thermoerosion dynamics in the design
Mining industry	Thermal effect of water on ice-containing rocks	Formation of coastal placers of mineral resources	
Water supply	Thermal effect of water on ice-containing rocks	Chemical contamination of water during destruction of woody and peaty coasts	
Crop raising, grassland farming	Thermal effect of water on ice-containing rocks	Destruction of arable and grazing lands	Redirection of erosive flow to the opposite shore
Crop raising	Breakdown of lands as a result of ravine thermoerosion	Complication of agricultural equipment operation, loss of agricultural lands	Backfilling of ravines, changes of boundaries and structure of agricultural lands
Automobile and railway transport	Breakdown of lands as a result of ravine thermoerosion	Erosion of roadbed, increase of the length of roads due to construction of bypass routes	Construction of water collectors and water-leading structures
Automobile transport	Removal of ground particles, resulting in eroding of tracks	Complication of road operation, increased depreciation of transport	Construction of hard-surfaced roads, auto service

The cracks are generally parallel with each other, and the same system includes cracks formed at right angles to them (Figure 28). In spring, snowmelt flows into these cracks and freezes there. When this process is repeated many times, a system of cavern-load ice is formed.

4.5.2. Impacts on engineering structures

Frost cracking constitutes a certain danger for the following engineering **structures**: (1) motor roads (roadways may go over the discontinuity); (2) residential and industrial buildings (breakage of continuous footings, cracks in the walls); (3) airfields (damage to airfield pavements); (4) pipelines (deformations and even breaks of underground steel pipelines); and (5) underground communication cables.

The scientific literature contains many descriptions of the effects of cryogenic cracking in **roadway coverings** (Figure 29). For example, systems of parallel cracks spaced 3 to 16 m apart develop in November, in sandy-loam grounds on roads with cement-ground covering in the Omsk Region of Russia. Their widths reach 7–25 mm in January through February, while, in summer, the cracks practically close. A similar case has been described for the city of Svobodny (Amur Oblast, Russia). On an earth roadbed constructed on sandy ground, transversal cracks measuring 40–50 cm deep and 1–3 cm wide were observed early in November at temperatures of –25° to –30°C; the distances between them reached 10–35 m (Grechishchev et al. 1980). Similar cases also have been noted on a motor road constructed on asphalted sand over a loamy foundation in the city of Yakutsk (Geocryological Dangers 2000).

Examples of influences of cryogenic cracking on **industrial and residential structures** are also fairly numerous. In one case, cryogenic cracks developed in a strip building foundation in Chita (Russia) due to a great drop in temperature to –27°C. Two cracks measuring 8–15 mm wide and 4.2 m apart got through the brick foundation, disrupting it (Foundations of Geocryology 2001, Vol. 4). In another case, a crack that opened up to 8 cm formed in a reinforced-concrete spandrel beam of the monolithic socle flooring and in the wall of a four-story house in Yakutsk (Grechishchev et al. 1980). Frost cracks also form in construction pits when buildings are constructed in winter. They are characteristic of practically all construction sites in regions of the High North (Geocryological Dangers 2000).

The formation of cryogenic cracks creates serious problems during the use of flight strips at airports. For example, the long-term development of frost cracks in asphalt coverings at an airport in the town of Amderma (Russia) necessitated its reconstruction in the early 1990s (Foundations of Geocryology 2001, Vol. 4).

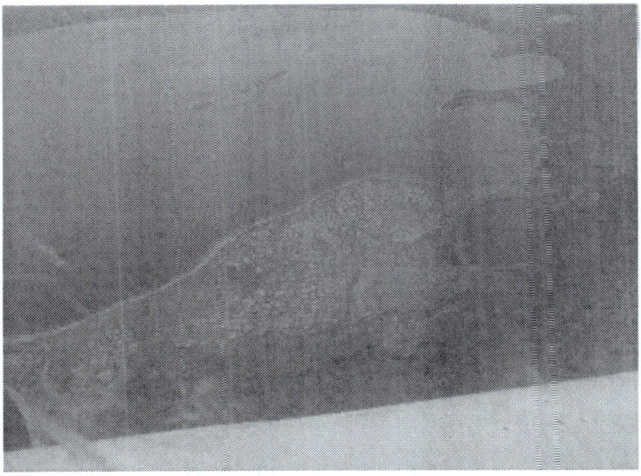

Photograph credit: S.P. Davydov (North-East Scientific Station, Pacific Geographical Institute, Cherskiy, Russia), July 2007.

Figure 28. Cryogenic cracking is generated by stretching strains emerging in frozen ground. In spring, water from melting snow penetrates into the ground and freezes therein. Repetition of the process leads to cavern-load ice formation. The photo shows polygon net near Cherskiy settlement (downstream of Kolyma River, Russia).

Photograph credit: S.Y. Parmuzin, Department of Geocryology, Moscow State University, 1967.

Figure 29. Cryogenic cracking oftentimes create problems for auto-road and railroad exploitation. The photo shows frost-induced cracks which deform roadbed in Zabaikalye.

There are *known* cases of considerable deformations and even ruptures of steel underground **pipelines** as a result of interactions with cryogenic cracks (Geocryological Dangers 2000). The full-scale experiments of I. N. Votyakov and G. P. Kuzmin (Grechishchev et al. 1980) showed that, where a pipeline axis laying depth is 0.6 m, the pulling stresses in an underground pipeline measuring 529 mm in diameter reach 27 x 10^3 kgf early in February (when the maximum opening of frost cracks occurs). Cryogenic cracking created serious problems for the construction and operation of the Tas-Tumus – Bestyakh gas pipeline. The effects on **underground communications cables** are similar to those on pipelines.

Measures to mitigate the effects of cryogenic cracking include only snow amelioration and consideration of shearing and pulling stresses arising in the frozen rocks and structures.

The influences of cryogenic cracking on human activities are outlined in Table 8.

4.6. Solifluction

Solifluction is a slow viscous-plastic flow of thawing waterlogged soils and fine-dispersed ground on gentle slopes. It occurs in Russia (Chukotka, Yana-Kolyma plain, Polar Urals, mountains of Siberia), the United States (Alaska), Canada, Norway (especially on the Svalbard Islands), the Falkland Islands, and mountainous regions of Central Asia.

4.6.1. Kinds of solifluction

The **conditions** necessary for the development of solifluction include the following (Romanovsky 1993): (1) increased content of pulverescent particles; (2) increased humidity; (3) presence of surface slopes (usually 2–3 to 10–15 deg); and (4) absence of woody and large shrub vegetation.

A **distinction** is made between mantled and differential (Figure 30) solifluction. For the former, relative areal uniformity, low drift velocity (0.5–10 cm/yr), and an absence of sinter relief forms are characteristic. For instance, surface solifluction rates at Steinhoi, Dovrefjell, Norway, over the period 2002–2006 ranged from 0.5 cm/yr at the rear of the lobe tread to 1.6 cm/yr (Harris et al. 2008). The distinctive feature of *differential solifluction* is the presence of characteristic forms of micro- and mesorelief: solifluction "tongues," flows, strips, terraces, etc. Their formation is caused by differences in drift velocities of thawing rocks on different parts of a slope. The rate of this type of solifluction may reach 10 cm/d (The Unquiet Landscape 1981). The areas of the solifluction relief forms range from several to thousands of square meters (Kaplina 1965).

One kind of solifluction is the *slip-out* (so-called fast solifluction). It is characteristic of steeper slopes (not less than 10 deg) formed by silt sandy loams or clay loams; fast solifluction has a catastrophic character but develops within relatively small areas. In the case of fast solifluction, rates reach tens of meters per day (Kozlova et al. 1992).

Solifluctional slip-outs (Figure 31) are prevalent on the Gydan Peninsula and Yamal Peninsula (Russia). In some places, this process affects up to 60–75% of the area of the slope. The volume of transported rocks in such regions reaches 35,000 m^3/km^2, while the denudation rate is 0.3 mm/yr (Geoecology of the North 1992). In one sections of the Yana-Omoloy interstream area, 2.25 million m^3 of thawing rocks were carried away from the slopes with area of 0.55 km^2 over a period of 3 years; in this case, the rocks proper accounted for 0.47 million m^3 while, on the ice, 1.78 million m^3 fell (Geocryological Dangers 2000).

4.6.2. Impacts on engineering structures

The influences of fast and slow solifluction are most urgent for the following **kinds** of human activity: (1) mineral resource industry; (2) transport (motor, rail, pipeline); and (3) industrial and civil engineering.

Table 8. Effects of cryogenic cracking on infrastructure

Basic objects	Nature of the effect	Consequences of the effect	Mitigation measures
Motor roads, airports	Shearing and pulling stresses arising in frozen rocks and resulting in cracking	Damage to road and airstrip coverings, increased depreciation of transport	Thermal and water insulation, replacement of dispersed grounds with gravel and sand mixtures
Industrial and civil development, pipelines	Shearing and pulling stresses arising in frozen rocks and resulting in cracking	Damage to buildings and structures (rupture of strip foundations, cracks in the walls, etc.)	Snow amelioration, consideration of possible stresses in structures, biological recultivation

A negative influence on the **mineral resource industry** is expressed as the *complication of operation* of enterprises due to sloughing of pit walls. Another consequence is *dilution* (reduction in concentrations of the commercial component). During mining operations, rocks containing the commercial component are stored in certain places for the purpose of downstream processing. Grounds that move under the action of solifluction increase the volume of rocks requiring processing, which reduces the economic efficiency of the operation of a mining enterprise.

Photograph credit: U.S. National Geophysical Data Center.

Figure 30. Feature of differential solifluction is generation of micro- and mezolandforms that are conditioned by different velocities of shifting of melting ground on different spots of the slope. At times, the speed of this kind of solifluction can amount 10 cm/day, but customarily it does not exceed 10 cm/year. The photo shows solifluctional flows near Suslositna Creek, Alaska.

Photograph credit: V.E. Tumskoy, Department of Geocryology, Moscow State University, Russia.

Figure 31. The slipouts (so called fast solifluction) are one of the kinds of solifluction. It is characteristic of the steeper slopes formed by silty loams or sandy clays. Its rate reaches several tens meters per day. Photo shows the solifluction slipout on the bank slope in Yakutia, Russia.

Photograph credit: Rejean Couture, Canada Geological Survey.

Figure 32. Solifluction has a certain positive importance for the transportation of heavy minerals to the valleys of rivers and streams and the formation of placer mineral deposits. The photo shows the solifluctial earthflows on the slopes of the Mackenzie River (Alaska, USA).

This problem is quite acute in the Yana-Omoloy interstream area. The intensive development of soliflunctional slip-outs results in the continual dilution of metallic rocks stocked on altiplanation terraces. In order to **deal** with this problem, protective fences are constructed there; however, they are frequently broken under the pressure of masses of transported ground (Foundations of Geocryology 2001, Vol. 4).

At the same time, slow solifluction has a certain **positive importance** for the transportation of heavy minerals to the valleys of rivers and streams and the formation of **placer mineral deposits** (Figure 32).

The effects on **transport** lie, first of all, in the deformation of hollows in the bodies of motor roads and railroads, and complications in the operation of surface pipelines. Soliflual slip-outs on the slopes of railway cuttings have repeatedly complicated the maintenance of the Novokuznetsk-Barnaul, Kozelsk-Plekhanovo, and Taishet-Lena railway lines in Russia (Kaplina 1965).

Table 9. Effects of solifluction on infrastructure

Basic objects	Nature of the effect	Consequences of the effect	Mitigation measures
Mining enterprises, motor and railway roads, ground-surface pipelines	Dynamic effects of transferrable material in the course of solifluction	Complication of operation of objects (deformation of road embankments, dulling of quarry borders and roadway excavation)	Fastening of the surface ground layer with vegetation, thermal insulation, drainage
Mining industry	Overlapping by layer of deposits as a result of soliflunc-tional slip-outs	Lowering of economic effective-ness of production due to dilution of warehoused rocks	Construction of barriers, fastening of the surface ground layer with vegetation, drainage
Mining industry	Transportation of heavy minerals to valleys in the course of slow solifluction	Formation of placer mineral deposits	

Problems for **industrial and civil engineering** are similar and consist mainly of sloughing of construction pit walls. It must be noted that the detrimental effects of solifluction are not infrequently intensified by human activities (destruction of topsoil, disturbance of ground temperatures, and others). As for **control** measures, water-thermal amelioration, binding stabilization of the surface ground layer with vegetation, and restriction of traffic flow with space and time are some recommended methods.

The effects of solifluction on human activities are outlined in Table 9.

4.7. Rock Streams

Rock streams are large-scale accumulations of large rock debris that cover mountain slopes (Figure 33) and flattops. Among the countries with **widespread occurrence of rock streams** are Russia, Canada, the United States, and Norway. They are also encountered in Sweden, Finland, New Zealand, mountainous regions of Italy, France, India, China, Chile, and other countries.

Russia shows a maximum distribution of rock streams. They occur widely in the mountain regions of Siberia and the Far East (for example, they occupy about 4% of the area of Transbaikalia). Rock streams occur most commonly within a zone between 54° and 64°N, and they decrease north and south of this zone (Foundations of Geocryology 2001, Vol. 4).

To the north of this zone, the lack of rock streams is related to decreases in outcroppings of rock-stream-forming rocks and decreases in the depths of seasonal thawing. To the south of this zone, the formation of rock streams is reduced due to decreases in the area of perennially frozen rocks. As one travels westward, the areas occupied by rock streams decrease drastically due to decreases in climate severity and decreases in elevation(Rock Streams of the Goltsy Altitudinal Belt 1989).

The **conditions** necessary for rock stream formation are as follows (Govorushko 1986): (1) the presence of rocks that generate coarse-grained material in the course of weathering; (2) climatic conditions that contribute to this process; (3) the presence of a slope with steepness not exceeding the angle of rest; and (4) the occurrence of bedrock close to the surface.

Rock streams also can occur where not all usual conditions that lead to them are present. In this case, the rock streams are relict and, for the most part, they formed during the period of the last climatic cooling, about 12,000 to 100,000 years ago (Rock Streams of the North Transbaikalia 1992).

Photograph credit: S.M. Govorushko (Pacific Geographical Institute, Vladivostok, Russia), August 1982.

Figure 33. Rock streams are large-scale accumulations of large rock debris that cover mountain slopes and flat tops. Photo shows rock stream in Myao-Chan Range (Far East of Russia).

As for the **feed sources,** two **types** of rock streams have been identified. The first type includes rock streams with *external* sources of feeding. They are formed by gravity-induced processes (rockfalls, screes) at the bottoms of scarp slopes. For rock streams with *internal* feeding, the macrofragmental material is obtained from bedrock subjected to weathering.

4.7.1. Mechanisms of rock streams

The spatial displacement of rock stream material results from the slope (surface) movement of fragmental product and suffosion of silt under rock streams. The **reasons** leading to rock stream **movement** vary, but, more often, they are different kinds of **creep**: *cryogenic* (displacement due to changes in deposit volume in the course of water freezing and ice melting), *thermogenic* (changes in the volume of fragments due to temperature variations), *hydrogenous* (changes in volume of silt due to changes in humidity), *suffosion*, and *plastic deformations of the ice-ground layer* (Govorushko 1986).

The **speeds** rock streams move generally range from millimeters to some centimeters a year, but sometimes there are catastrophic motions caused by different reasons (earthquakes, sharp ice thawing the rock stream mass during abnormally warm summers, etc.).

4.7.2. Impacts on engineering structures

In the most general case, the stability of engineering projects on the slopes of rock streams decreases with increases in the steepness of slopes, thicknesses of goltsy ice and layers of inundated thixotropic fine-grained soil in the base of the rock stream, and other factors The main reasons rock stream cover becomes destabilized are (1) cutting of slopes, which removes the rock mantle "backstop"; and (2) changes in subsurface flow when mounds are partitioned and the voids are filled with fine earth.

Rock streams affect the following **kinds of human activity and structures**: (1) motor transport; (2) rail transport; (3) hydraulic engineering construction; (4) mineral resource industry; (5) search for mineral deposits; (6) grassland farming; (7) populated localities; (8) building materials industry; and (9) water supply systems.

Photograph credit: A.I. Tyurin (Department of Geocryology, Moscow State University, Russia), 1987.

Figure 34. The rock streams oftentimes aggravate the construction and maintenance of railroads and auto-roads, hydraulic structures, etc. The photo shows the rock stream crawling upon the road near the Udokan mountain ridge, water basin of the Naminga River (Russia).

The effects on **motor roads** include pressure of the coarse-grained mass on road embankments (Figure 34), blockage of roadbeds with deposits, and washout of embankments by water discharge under rock streams, among others.

The cases of construction of motor roads through rock streams are not numerous. They have been built in the southern Urals (Russia); and V. S. Fedotov (1966) points out that, at points where they cross rock streams, the roads deteriorate faster. Year after year, the Mogocha-Chara motor road needs to be cleared of rubble, and on one of the temporary motor roads in the South Muya ridge, the motion of rock streams destroyed retaining walls. During construction of a motor road on a rock stream at the Nyukzha River (tributary of Olekma River, Lena River basin), the bed was washed away in the summer and ice crust formed during autumn and winter (Gordeev et al. 1980). Similar problems arose in a number of sections of the Amur-Yakut arterial road (Geocryological Dangers 2000).

Impacts on **rail transport** are similar to those on motor transport. In cases of disastrous movements, the consequences may be very serious. An example of such an occurrence is a freight train accident in the Severomuisky section of the Baikal-Amur Railroad (Russia) in 1990 (Exogenous Geological Hazards 2002). In all other cases, slow deformations of embankments usually occur. By now, a certain amount of experience in railway construction on rock streams has been gained. Nearly a century ago, the Circum-Baikal Railway was constructed and, afterwards, the Novokuznetsk-Abakan, Askiz-Abakan, Abakan-Taishet, Bystrovka-Rybachye, and Mezhdurechensk-Abakan railways were erected (Nikitenko 1970). At the present time, 4,234 m of railway exist on dynamic rock streams in the Kuznetsky Alatau (Glazovsky 1978).

During construction of the Baikal-Amur Mainline (BAML), the construction workers were confronted with considerable difficulties laying the earth roadbed on rock streams. In most cases, the rock streams were bypassed. The body of railroad was transferred to the opposite side of the river, which required that additional bridges be built; all of this added to the costs. The construction of the railway around the South Muya ridge proved to be impossible. Thick layers of ice were present below the continuous cover of rock streams. Cutting the rudaceous cover would cause the ice to melt, which would unavoidably result in collapse of the railway (Geocryological Dangers 2000). Thus the attempt to fully avoid construction along the rock streams failed. Sections that were constructed over rock streams were generally confined to the central part between the Chara and Tynda stations. A detailed description of methods of BAML construction on rock streams can be found in a monograph by S. M. Govorushko (1986).

Effects of rock streams on the volume of **water storage reservoirs** happen during the course of two processes of different directionality (Govorushko 2008). The *first* process is a reduction in volume because a basin fills up with fragmental products and silt. The *second* is an increase in the basin volume when icy filling material melts out of flooded rock streams and the coarse-grained material subsides. Investigations carried out in the water storage reservoir of the Kolyma hydropower station (Russia) showed that, for its possible operating period, the effect of rock streams has been to increase the volume of the storage reservoir basin (Govorushko 1984). For the most part, the effects of rock streams in changing the volumes of storage reservoirs are insignificant, as is the intensity of erosion of shorelines formed by rock streams.

Rock streams have certain effects on **mining enterprises**. In addition to complicating the construction of the mines themselves, they have some impact on the volumes of *tailing dumps*—spots where barren rock that may be reworked in the future is accumulated. In the majority of instances, the effects of rock streams on tailing dumps result in an insignificant reduction in their volume (Govorushko 2007).

The construction principles related to the effects of rock streams on tailing dumps are similar to those for water reservoirs. The major relevant characteristic of a tailing dump is the capacity of its basin. The influence of rock streams depends on their mobility and the amount of fine-grained soil that is removed. A study of the dynamics of rock streams in the Miao-Chan ridge led to the conclusion that they have a minimal effect on the tailing dump of the mining and processing works in the Gorny settlement (Khabarovsk Krai, Russia) (Govorushko 1986). This is likely to be typical of similar situations.

Where they are present, rock streams leave traces **of mineral deposits**. On the one hand, they create difficulties in the application of some search techniques (for example, electrical exploration), while, on the other hand, the peculiarities of rock stream dynamics can simplify the detection of mineral deposits. One widely accepted method of detection is the lithochemical survey. The ore bodies form a so-called dispersion halo when the surface layers are enriched in the commercial element. According to a specific grid, the samples are taken and analyzed to detect the desired deposits. As a rule, the ore bodies on rock stream slopes have closed dispersion halos, and shallow sampling does not detect the deposits.

However, suffosion under rock streams can selectively wash out fine ore matter from mineralized zones. As a result, material enriched with ore elements is carried out to the foot of a slope. A technique of searching for mineral deposits on slopes with rock streams is based on this. In valleys, samples of silt carried out from under rock streams are taken, and, based on increases in the content of useful components, ore bodies are found on the overlying slopes (Taisayev 1981). This technique is suitable for discovery of deposits of molybdenum, nickel, copper, tin, tungsten, and other metals (Govorushko 1986).

To some extent, rock streams have impacts on **grassland farming**. They can destroy a soil layer, and grass cover, shrubs, and woody vegetation die out. Rock streams create problems for reindeer breeders: when a herd of deer crosses a rock stream, animals often break their legs (Geocryological Dangers 2000). Similar injuries resulting in death are also suffered by wild deer and elk, especially in winter when debris is below the snow. To facilitate the travel of reindeer, breeders construct wooden tracks on the rock streams (Figure 35), which are wrecked fairly quickly (Geocryological Dangers 2000).

For the present, there is no need to use territories where rock streams are present for purposes of **residential building**, although in the neighborhood of the towns of Zlatoust and Katav-Ivanovsk (Russia), rock streams approached the domestic buildings, and, in a number of villages of the South Ural (Russia), they negatively impacted truck farming (Fedotov 1966). Similar situations also occurred in a number of settlements along the Baikal-Amur Mainline, where the tongues of rock streams were located between buildings (Geocryological Dangers, 2000).

Rock streams can be used to supply *building materials*, *broken-stone ballast*, and, in certain cases, decorative facing material. They played an essential positive role in the construction of the Baikal-Amur Railroad, when coarse deposits of rock streams were used to construct the roadbed. In addition, ice present in rock streams has been used for **local water supplies,** which we repeatedly observed near the Verkhne-Kolyma range and the Dzhugdzhur ridge (Govorushko 2008).

In the construction of commercial facilities, it is necessary to prevent such impacts on the rock streams as cutting of slopes and changes in subsurface runoff. These goals are attained by methods such as allowing for subsurface runoff drainage, creating water sluices, preventing the formation of hollows in the rock streams (at least, side-hill fill), and keeping embankment material from penetrating the body of the rock stream. Because the railway embankments are displaced together with a rock stream, M. K. Druzhinin

(1987) recommends that embankments be wider on the upstream side so that railway tracks can be realigned while they are in operation.

Generally, one can say that the principles of construction on rock streams have been developed mainly theoretically, and they have not been observed in practice to a large extent, as engineering development on rock stream slopes is not yet appreciable.

The effects of rock streams on human activities are outlined in Table 10.

Photograph credit: A.I. Tyurin (Department of Geocryology, Moscow State University, Russia), 1989.

Figure 35. Rock streams create problems for reindeer breeders: when a herd of deer crosses a rock stream, animals often break their legs. The photo shows the wood-strip path in the rock stream designed for pass of deers (Vitim River basin, Russia).

Table 10. Effects of rock streams on infrastructure

Basic objects	Nature of the effect	Consequences of the effect	Mitigation measures
Motor and rail roads	Outflow of water of the seasonal-thawed layer and aufeis formation on the roadbed in case of rock stream cutting	Blocking roads by aufeis, difficulties in traffic flow, interruptions in road operation	Embankment fill using crushed stony material and collecting ditches for interception of water, filling of cavities in the block foundation with polystyrene foam
Motor and rail roads	Melting of goltsy ice in the crests of artificial slopes, removal of fine-grained soil below roadbed	Deformation of the roadbed due to thermokarst subsidence and suffosion, difficulties in road operation	Creation of water sluices under the roadbed with protection against penetration of debris into them
Motor and rail roads	Catastrophic displacements of detrital material due to earthquakes, heavy showers, etc.	Destruction of roadbed	Bypass of dangerous areas, non-use of cuttings of rock streams, construction of galleries
Motor and rail roads	Displacement of detrital material in linear rock streams	Destruction of roadbed	Construction of bridge crossings, bypass of dangerous areas
Rail roads	Displacement of detrital material	Displacement of embankment downslope, deformation of railway track	Widening of the embankment and track alignment for straightening of slippage curve, construction of trestles
Mining enterprises	Slope displacement of detrital material and removal of fine-grained soil into the tailing dump basin	Reduction in usable storage of tailing dumps	Adequate choice of site and normal headwater level based on examination of natural conditions
Search for mineral deposits	Formation of secondary dispersion halo of ore bodies for removal of fine-grained soil below the rock stream	Discovery of mineral deposits by testing transported material	
Single construction works (pumping	Displacement of detrital material	Foundation destabilization, deformation of structures	Choice of reliable foundations (rocky residual outcrop)

Basic objects	Nature of the effect	Consequences of the effect	Mitigation measures
houses, filling stations, etc.)			
Hydroelectric power stations	Slope displacement of detrital material, ice spreading, removal of fine-grained soil into the water reservoir basin	Reduction in usable storage of water reservoirs	Adequate choice of site and normal headwater level based on examination of natural conditions
Hydroelectric power stations	Melting of goltsy ice contained in flooded rock streams	Increase in capacity of water reservoirs	
Grassland farming	Problem of run of deer through rock streams due to abundance of cavities	Death and trauma of animals as a result of falls between blocks	Use of bypass routes, construction of wood-strip paths for run of animals
Production of construction materials	Formation of rock streams due to weathering of parent rocks and gravitational processes	Use of rock stream deposits as construction material, broken-stone ballast	

4.8. Rock Glaciers

A **rock glacier** is an accumulation of rock material cemented (consolidated) with ice. Rock glaciers are characteristic of many mountain systems with cold and moderately humid climates (Figure 36). They are **abundant** in the Alps, Caucasus, Pamirs, Hindu Kush, Tien Shan, Karakoram, Rocky, and Andes Mountains, and in Greenland, Spitsbergen, New Zealand, and coastal areas of Antarctica (Gorbunov 2008). Small rock glaciers also occur in the high mountain areas of Africa and Central America.

Disko Island, off the western coast of Greenland, can be called a real **kingdom of the rock glaciers**. Its area is 8,575 km^2, and its maximum height reaches 1,904 m. This island has about 1,700 rock glaciers (Gorbunov 2008). In northeastern Russia, with an area of 1.57 million km^2, there are 6,500 rock glaciers (Galanin 2009). In the Swiss Alps, 994 active rock glaciers have been identified (Barsch 1996).

For the earth as a whole, areas occupied by rock glaciers are much smaller as compared with those occupied by glaciers. With respect to glaciers, rock glaciers occupy a *lower hypsometric position* (Galanin 2005).

4.8.1. Types of rock glaciers

Two categories of rock glaciers have been identified: (1) ice-cemented rock glaciers (cryogenic formations having no historico-genetic relation with glaciers) and (2) ice-cored rock glaciers (formed from glaciers in the course of their reduction and burial under layers of detritus).

Ice-cored rock glaciers are more numerous, and sometimes it is quite difficult to distinguish between a glacier and a rock glacier. For example, based on detailed study of structure, mass exchange, and dynamics, many glaciers in the Alps have been determined to be rock glaciers (Krainer et al. 2002).

According to dynamic activity, the following **types** of rock glaciers have been identified: (1) active; (2) inactive; and (3) fossil rock glaciers (Barsch 1996). The drift **speeds** of active rock glaciers are, on average, several tens of centimeters per year. Speeds of some meters per year have been recorded in the Andes where there are abrupt increases in slope inclination (Kotlyakov 1994). Under certain conditions (for example, at seismic points), catastrophically fast drift of vast masses of ice-rock material is possible, such as that observed in the course of the Yamsky earthquake of 1851 in Priokhotye (Russia) (Galanin 2005).

Photograph credit: http://en.wikipedia.org/wiki/Rock_glacier.

Figure 36. Rock glaciers are distinctive geomorphologic landforms of angular rock debris frozen in interstitial ice which may extend outward and downslope from scree cones, glaciers, etc. Photo shows rock glacier with multiple flow lobes, Chugach Mountains, Alaska.

As for the *sizes of rock glaciers*, their **lengths** vary from hundreds of meters to 3 km (http://en.wikipedia.org/wiki/Rock_glacier), **widths** reach hundreds of meters and even kilometers, while the maximum **thickness** is about 100 m (Galanin 2008). Rock glaciers end in benches with heights of 15–50 m and sometimes up to 80–100 m.

4.8.2. Impacts on engineering structures

Rock glaciers influence chiefly the following **kinds of human activity**: (1) transport; (2) water power engineering; (3) residential construction; (4) the mineral resource industry; and (5) water supplies.

The effects on **motor and rail transport** are related to the dynamic action of the rock-ice mass and blockage of roads with deposits (Figure 37). The influence on **water power engineering** is generally *positive* because of the considerable contribution of rock glaciers to the nourishment of rivers on which hydropower stations have been constructed.

The effects on **residential construction** are caused by both breaks of rock glacier–dammed lakes and their slow drift when they destroy houses and other structures. Examples of both types of events for the southern Chukot Peninsula are presented by A. A. Galanin (2005). He notes a great lake dam near the settlement of Provideniya, as well as the danger posed by rock glaciers to several buildings in the settlement of Ureliki.

The effects of rock glaciers on the **mineral resource industry** include creating difficulties for mining operations through the blockage of corrals, filling of open pits with rock-ice masses, and other problems. Mining projects also affect rock glaciers, so there is interaction between them. For instance, 26 mining projects have already affected rock glaciers in Chile, Argentina, and Peru; the affected rock glacier area in Chile is approximately 3.3 km^2 (Brenning and Azocar 2010). Also, rock glaciers in the Chilean Andes help supply the water for much of Chile, including the capital, Santiago (http://en.wikipedia.org/wiki/Rock_glacier).

The effects of rock glaciers on human activities are outlined in Table 11.

Photograph credit: S.M. Govorushko (Pacific Geographical Institute, Vladivostok, Russia), August 28, 2006.

Figure 37. The impact of rock glaciers on human activity generally is not great. The photo shows the threatening movement of a rock glacier towards the Fluela pass of the road connecting Switzerland and Italy. This pass is the watershed between the basins of the Rhine and the Danube Rivers.

Table 11. Effects of rock glaciers on infrastructure

Basic objects	Nature of the effect	Consequences of the effect	Mitigation measures
Residential and industrial development	Dynamic effect of ice mass as the rock glacier advances	Destruction of buildings and structures	Transfer to the other location, considering the dynamics of rock glaciers in construction
Automobile transport	Blockage of roadbed with deposits of rock glacier	Destruction of roadbed, blockage of traffic	Bypass of dangerous areas when constructing roads
Crop raising, cattle breeding, hydraulic power industry	Contribution of waters during thawing of rock glacier to river feeding	Possibility to use waters for irrigation needs, increase in electric energy generation	
Hydraulic power industry	Water displacement when rock glacier slides down to the water reservoir	Reduction in usable storage of the water reservoir, dam failure	Well-timed water discharge in order to avoid water overflow through the dam
Mining industry	Burial of orifices of mines when rock glaciers are in motion	Complication in development of deposits, economic damage	Consideration of the dynamics of rock glaciers in developing mines and mine galleries
Water supply	Melting of ice of rock glaciers	Use of melt water as water supply for populated localities	

CONCLUSION

This chapter has shown how cryogenic processes substantially complicate human activities in the northern regions. For the time being, the economic damage they cause is no too high in view of the poor development of territories with perennially frozen rocks. However, development is likely to intensify due to extraction of commercial minerals (petroleum, gas, gold, etc.), which will enhance the negative significance of their impact. Thus further study of cryogenic processes is necessary to develop measures of protection against and mitigation of their effects.

References

Abaturova, IV; Gryaznov, ON; Storozhenko, LA; Yemelyanova, IA. Present-day natural and techno-natural exogenous geological processes of the Ob-Nadym interfluve. Geoecology. *Engineering Geology, Hydrogeology, Geocryology*, 2009, No. 2, 161–171 (in Russian).

Anderson, GK; Trigg, KF. *Interesting Cases from the Practice of Engineering Geology*; Nedra: Moscow, 1981, 224 pp. (in Russian).

Are, FE. *Principles of the Shore Thermoabrasion Forecast*; Nauka: Novosibirsk, 1985, 172 pp. (in Russian).

Atlas of Natural and Technogeneous Dangers and Risks of Emergency Situations in the Russian Federation; IPTS, *Design. Information. Cartography*: Moscow, 2005, 271 pp. (in Russian).

Barsch, D. Rock Glaciers: *Indicators for the Present and Former Geoecology in High Mountain Environments*; Springer-Verlag: Berlin, 1996, 331 pp.

Brenning, A; Azocar, GF. *Mining and rock glaciers: environmental impacts, political and legal situation, and future perspectives*. Revista De Geografia Norte Grande, 2010, Is. 47, 143–158.

Brown, J; Jorgenson, J. Index plots of Beaufort Sea in Alaska as the part of the program of the Arctic coastal dynamics monitoring. *Extreme cryospheric phenomena: fundamental and applied aspects*. Publ. PSC of RAS: Pushchino, 2002, 60–61.

Druzhinin, MK. *Principles of Engineering Geology*; Nedra: Moscow, 1987, 223 pp. (in Russian).

Exogenous Geological Hazards; Kruk: Moscow, 2002, 348 pp. (in Russian)

Fedotov, VS. Rock streams and taluses of the Southern Urals. In: *Problems of Geology and Geomorphology of West Siberia*. Book publisher: Barnaul, 1966, 18–36 (in Russian).

Foundations of Geocryology. Vol. 4. *Dynamic Geocryology*. Yershov, E. D., Ed. Moscow State University: Moscow, 2001, 688 pp. (in Russian).

Galanin, AA. Rock glaciers are special type of modern mountain glaciation in northeastern Asia. *Vestnik of FEBRAS*, 2005, Vol. 5, 59–71 (in Russian).

Galanin, AA. Rock glaciers of northeastern Asia: mapping and geographical analysis. *Cryosphere of Earth*, 2009, Vol. XIII, No. 4, 49–61 (in Russian).

Galanin, AA. Rock glaciers: history of study and present-day ideas. Vestnik of the North-Eastern Scientific Centre of FEB RAS, 2008, Vol. 3, 15–33 (in Russian).

Geocryological Dangers. *Natural Dangers of Russia*, Vol. 4; Izdatelsky firma KRUK: Moscow, 2000, 316 pp. (in Russian).

Geoecology of the North (*Introduction in Geocryology*); Publ. House of Moscow University: Moscow, 1992, 270 pp. (in Russian).

Gevorkyan, SG. Cryogenic natural processes in the cryolithic zone and preparation of the Arctic scene of war operations. International Conference, Geopolitics: theory, history, practice. In: *Theoretical and Methodological Background of Geopolitics and Globalistics*, Moscow, 2012, 70–77 (in Russian).

Glaciological Encyclopaedia; Gidrometeoizdat: Leningrad, 1984, 528 pp. (in Russian).

Glazovsky, AF. Rock glaciers (state of the problem). In.: *Cryogenic Phenomena of Highlands*; Nauka: Novosibirsk, 1978, 59–72 (in Russian).

Gorbunov, A. P. The rock glaciers of the world: review (Report 2). *Cryosphere of the Earth*, 2008, Vol. 12, No. 3, 58–68 (in Russian).

Gordeev, PP; Murzin, Yu A; Goncharov, VK. Rock streams of the middle Olekma basin. Geocryologic conditions of the Baikal-Amur Mainline zone. IG SB of the USSR Academy of Sciences: Yakutsk, 1980, 103–112 (in Russian).

Govorushko, SM. Forecast of effect of rock streams on the capacity of water reservoirs. *Geomorfologiya*, 1984, No. 2, 37–42 (in Russian).

Govorushko, SM. *Impact of Geologic, Geomorphologic, Meteorological and Hydrological Processes on Human Activity*; Academicheskiy proekt: Moscow, 2007, 684 pp. (in Russian).

Govorushko, SM. *Importance of rock streams for economic activity. Problems of Environment and Natural Resources*, 2008, Vol. 7, 95–100 (in Russian).

Govorushko, SM. *Morpholithogenesis of rock streams*. Far Eastern Scientific Center of the USSR Academy of Sciences: Vladivostok, 1986, 120 pp. (in Russian).

Govorushko, SM. Natural Processes and Human Impacts. *Interactions between Humanity and the Environment*; Springer: Dordrecht, 2012, 678 pp.

Govorushko, SM. Underground ice and thermokarst in the lower reaches of Indigirka River. *Field and experimental studies of permafrost*. Institute of Permafrost Studies, SB of the USSR Academy of Sciences: Yakutsk, 1981, 34–39 (in Russian).

Grechishchev, SE; Chistotinov, LV; Shur, Yu L. *Cryogenic Physical and Geological Processes and Their Forecast*; Nedra: Moscow, 1980, 383 pp. (in Russian).

Grigor'ev MN; Khubberten, X -B; Razumov, CO. Cryogenic processes and dynamics of shores in the northwestern delta of the Lena River. *Extreme Cryospheric Phenomena: Fundamental and Applied Aspects; Publ. PSC of RAS*: Pushchino, 2002, 75–76 (in Russian).

Grigor'ev, MN; Schneider, W. Shore erosion processes and sediment flux from eroded islands in the apex of the Lena delta. Ber. Polar- und Meeresforsch, 2002, No. 426, 52–56.

Grosse, G; Jones, BM. Spatial distribution of pingos in northern Asia. *Cryosphere*, 2011, Vol. 5, Is. 1, 13–33; DOI: 10.5194/tc-5-13-2011.

Harris, C; Kern-Luetschg, M; Smith, F; Isaksen, K. Solifluction processes in an area of seasonal ground freezing, Dovrefjell, Norway. *Permafrost and Periglacial Processes*, 2008, Vol. 19, Is. 1, 31–47. http://en.wikipedia.org/wiki/Permafrost http://en.wikipedia.org/wiki/Rock_glacier http://www.britannica.com/EBchecked/topic/452187/permafrost/65728/Climatic-change

Ice that burns: gas hydrates offer a potential energy reserve, danger to petroleum exploration and global warming concern. *Research Horizons Magazine*, July 11, 2002; http://gtresearchnews.gatech.edu/newsrelease/HYDRATES.htm.

Isakov, V. Basic types and reasons of developing deformations on motor and rail roads in the Norilsk industrial district. *Proceedings of Tenth International Conference on Permafrost*, Vol. 3, Resources and Risks of Permafrost Areas in a Changing World; Pechatnik, Tyumen, 2012, 207–211 (in Russian).

Jones, BM; Grosse, G; Hinkel, KM; Arp, CD; Walker, S; Beck, RA; Galloway, JP. Assessment of pingo distribution and morphometry using an IfSAR derived digital surface model, western Arctic Coastal Plain, northern Alaska. *Geomorphology*, 2012, Vol. 138, Is. 1, 1–14; DOI: 10.1016/j.geomorph.2011.08.007.

Kamalov, AM; Ogorodov, SA; Arkhipov, VV. Dynamics of the West Yamal coasts, Extreme Cryospheric Phenomena: *Fundamental and Applied Aspects*; Publ. PSC of RAS: Pushchino, 2002, 63–64 (in Russian).

Kaplina, TN. *Cryogenic Slope Processes*; Nauka: Moscow, 1965, 295 pp. (in Russian).

Kokelj, SV; Jenkins, RE; Milburn, D; Burn, CR; Snow, N. The influence of thermokarst disturbance on the water quality of small upland lakes, Mackenzie Delta Region, Northwest Territories, Canada. *Permafrost and Periglacial Processes*, 2005, Vol. 16, Is. 4, 343–353.

Kondratyev, VG. Cryogenic risks and resources of the railroads in the cryolithic zone. *Proceedings of Tenth International Conference on Permafrost*, Vol. 3, Resources and Risks of Permafrost Areas in a Changing World; Pechatnik, Tyumen, 2012, 235–240 (in Russian).

Kotlov, FV. *Change in the Geological Environment under the Influence of Human Activity*; Nedra: Moscow, 1978, 263 pp. (in Russian).

Kotlyakov, VM. *World of Snow and Ice*; Nauka: Moscow, 1994, 248 pp. (in Russian).

Kozlova, AE; Nekrasova, LA; Sukhodrovsky, VL. Assessment of impact of cryogenic processes on pipelines of northern West Siberia. Engineering Geography. *Engineering-Geomorphologic Aspects*, Vologda, 1992, Is. 1, 62–64 (in Russian).

Krainer, K.; Mostler, W.; Span, N. A glacier-derived, ice-cored rock glacier in the Western Stubai Alps (Austria): evidence from ice exposures and ground penetrating radar investigation. Zeitschrift fur gletscherkund und glazialgeologie. Universitatsverlag Wagner, Innsbruck, 2002, Vol. 38, No. 1, 21–34 (in German).

Larionov, AK. *Entertaining Engineering Geology*; Nedra: Moscow, 1974; 2?0 pp. (in Russian).

Lolayev, AB. *Engineering-Geocryologic Problems of the Cryolithic Zone Development*; Moscow, 1998; 154 pp. (in Russian).

Lomtadze, VD. *Engineering Geodynamics*; Nedra: Leningrad, 1977; 479 pp. (in Russian).

Mackay, J. Ross. Pingo. Growth and collapse, Tuktoyaktuk Peninsula area, western Arctic Coast, Canada: a long-term field study. *Géographie physique et Quaternaire* (University of Montreal), 1998, Vol. 52, No. 3, p. 311; Retrieved June 23, 2012.

Malik, LK. Factors of Risk of Damage to Hydraulic Structures. *Safety Problems*; Nauka: Moscow, 2005, 354 pp. (in Russian).

Myagkov, SM. *Geography of Natural Risk*; Moscow University: Moscow, 1995, 222 pp. (in Russian).

Natural-Anthropogenic Processes and Environmental Risk. *Geography, Society, Environment*, Vol. IV; Gorodets: Moscow, 2004, 615 pp. (in Russian).

Nikitenko, FA. Ground Roadbed in the Sections of Developing Landslides, Rockslides and Rock Streams in the Mountain Regions of East Siberia. *Railroads under Severe Physico-Geographical Conditions*; Novosibirsk, 1970, 5–18 (in Russian).

Orlov, VO; Dubnov, Yu D; Merenkov, ND. *Heaving of Frozen Grounds and Its Effect on the Foundations of Structures*; Stroizdat: Leningrad, 1977, 183 pp. (in Russian).

Orlov, VO; Yelgin, BB; Jeleznyak, II. Frost *Heaving of Grounds and Its Calculations in Bases of Structures*; Nauka: Novosibirsk, 1987, 137 pp. (in Russian).

Popov, AI; Rozenbauem, GE; Tumel, NV. *Cryolithology*. Moscow University: Moscow, 1985, 239 pp. (in Russian).

Poznanin, VL. *Nature of ravine thermoerosion: Abstract of thesis of Dr. Sc. (geography)*; Moscow, 1995, 33 pp. (in Russian).

Rock Streams of the Goltsy Altitudinal Belt. Romanovsky, N. N.; Tyurin, A. I.; Sergeev, D. O.; et al.; Nauka: Novosibirsk, 1989, 152 pp. (in Russian).

Rock Streams of the North Transbaikalia; Nauka: Novosibirsk, 1992, 181 pp. (in Russian).

Romanovsky, NN. *Fundamentals of Lithosphere Cryogenesis*; Moscow State University: Moscow, 1993, 336 pp. (in Russian).

Russian Arctic: *On the Threshold of Catastrophe*; Center of the Ecological Policy of Russia: Moscow, 1996, 206 pp. (in Russian).

Stolbovoi, V. Carbon content in coastal ecosystems of Russian Arctic. Abstracts of International Workshop, Arctic Coastal Dynamics; Ber. Polar- und Meeresforch, 2002, No. 413, 75–76.

Sukhodrovsky, VL. *Exogenous Relief Forming in the Cryolithic Zone*; Nauka: Moscow, 1979, 280 pp. (in Russian).

Taisayev, TT. *Geochemistry of Taiga-Cryogenic Landscapes and Exploration of Ore Deposits*; Nauka: Novosibirsk, 1981, 137 pp. (in Russian).

The Unquiet Landscape; Mir: Moscow, 1981, 188 pp. (in Russian).

Theoretical Principles of Engineering Geology; Nedra: Moscow, 1985, 259 pp. (in Russian).

Titkov, SN; Ablyazina, DI. Impact of cryogenic soil heaving on stability of the Yamburg Gas-Condensate Field's objects. *Industrial and Civil Construction*, 2008, Vol. 11, 31–33 (in Russian).

Tomirdiaro, SV. Natural Processes and Development of the Permafrost Zone Territories; Nedra: Moscow, 1978, 145 pp. (in Russian).

Vasilyev, AA. Assessment of factors affecting the destruction of sea coasts in the western sector of the Russian Arctic. Extreme Cryospheric Phenomena: *Fundamental and Applied Aspects*; Publ. PSC of RAS: Pushchino, 2002, 73–74 (in Russian).

Vinogradova, O. Gas hydrates. Look to 2025. *Journal of Geocryology*, 2001, Vol. 3; http://www.netpilot.ca/geocryology/num3/Gashydro.html (in Russian).

Voskresensky, KS; Chaus, IK; Kholnov, AP; Romanenko, FA; Marakhtanov, VP. Appropriateness of the spreading and development of the modern exogenic processes on accumulating planes and lowlands of North of Russia—energetic method of approach. Information Bulletin of RFBR. *Earth Sciences*, 1999, No. 7, 34–43 (in Russian).

Vtyurin, BI. *Underground Ices of the USSR*; Nauka: Moscow, 1975, 214 pp. (in Russian).

Vtyurin, BI; Govorushko, SM. "Whims" of thermokarst. *Cryosphere of the Earth*, 2012, Vol. XVI, Is. 4, 42–44 (in Russian).

Vtyurina, EA; Vtyurin, BI. *Ice Formation in Rocks*; Nauka: Moscow, 1970, 280 pp. (in Russian).

Walker, HJ. Bluff erosion at Barrow and Wainwright, Arctic Alaska. *Z Geomorphol NF*, Suppl. Bd. 81, Berlin-Stuttgart, 1991, 53–61.

Walker, HJ. Coastal Processes and Their Influences on the People of Alaska's North Slope. *Fifty More Years Below Zero; Arctic Institute of North America: Calgary*, 2001, 117–122.

Walker, HJ; Arnborg, L; Peippo, J. Riverbank erosion in the Colville delta Alaska. *Geografiska Annaler*, 1987, Vol. 69A, No. 1, 61–70.

Walsh, MR; Hancock, SH; Wilson, SJ. Preliminary report on the commercial viability of gas production from natural gas hydrates. *Energy Economics*, 2009, Vol. 31, 815–823.

Yegurtsov, SA; Velikotsky, MA; Marakhtanov, VP; Chigir, VG; Menshikov, SN; Polozov, VN. *Dangerous cryogenic processes and operation reliability of the interfield trunk pipeline of the Medvezhye field.* Gas Industry, 2011, No. 2, 54–57 (in Russian).

Yershov, ED. *General Geocryology*; Publ. MSU: Moscow, 2002, 682 pp. (in Russian).

Yuryev, IV. Problems of operation of the gas complex objects within the littoral zone of the West Yamal. *Cryosphere of the Earth*, Vol. V, Vol. XIII, 2009, No. 1, 46–54 (in Russian).

In: Permafrost
Editor: Oleg S. Pokrovsky
ISBN: 978-1-62948-830-1

Chapter 2

PATTERNED GROUND AND CLIMATE CHANGE

B. Van Vliet-Lanoë*

Domaines Océaniques, University of Brest, CNRS, IUEM
Place N.Copernic, Plouzané, France

ABSTRACT

In memorian Link Washburn

Patterned ground is one of the commonest morphologies formed in various periglacial environments, with or without permafrost. This is also valid for polar, as well as mountain, environments. It is usually formed by differential frost heave on superposed sediments with various frost susceptibilities or the capacity to heave under the action of frost. They may develop as self-organised nets, merging as coalescing forms, or as cracking guided nets (thermal or desiccation), defining contiguous forms that are perennially stable. The frost susceptibility develops in sediments and soils in the form of ice lensing under control of the grain size, the freezing rate and the available moisture. With time, frost shattering, stone jacking by frost and particle leaching at thaw may modify the frost susceptibility, limiting or enhancing the capability for ice lensing. For moisture availability, patterned ground mostly develops on shallow permafrost, which retains a perched water-table in late summer, providing moisture for the winter frost. Liquefaction is a superficial and local process only. Any change in the annual thermal budget may modify the

* Corresponding Author address Email: brigitte.vanvliet-Lanoe@univ-brest.fr.

late-summer position of the water-table. With the recent warming, the permafrost-table usually becomes deeper, starving the active layer of available moisture for a differential frost heave; the patterned forms in former drained position inactivate; those in a former waterlogged position rise, deforming the layers. Lakes may drain, activating patterned ground on the freshly emerged bottom, but partly inactivating patterned ground fields located higher in the topography, as in Svalbard. If the precipitation is raised too, the changes are limited. If a cooling develops, as during the 1970's, the active layer is shallower but waterlogged in late summer: deformations are limited by a superficial rigidity (limited differential frost heave) due to the ground saturation in ice. Once initiated, the forms may survive for millennia if they are not disturbed by glacial extent or a marine transgression, adapting to the various thermal and moisture condition changes. Those forms may also develop on slopes, stretched by frost creep and evolving into ground stripes. Biological activity, especially in mountain areas may adapt to these change in drainage and frost heave, leading to specific vegetation morphologies. In this case, patterned ground may be partly biogenic as in temperate mountains.

INTRODUCTION

Patterned ground is one of the most common and beautiful morphologies formed in various periglacial environments. These landforms are associated with freezing and thawing on an annual or shorter-term basis, or with the seasonal frost, or permafrost or permanently frozen ground, which melts only superficially in summer. They are generally described as cryoturbations or involutions. As cold climates imply cold temperatures, at least in winter, water is present during a part of the year in the form of snow or ice. This is also valid for polar and mountain environments. As ice is needed for most of the specific mechanisms, water is the *sine qua non* of frost activity (Washburn, 1976). Ice formation and frost impact may be considered as a peculiar type of thermally induced desiccation as the vapour tension is low at a negative temperature.

Patterned ground morphologies cover "tundra polygons", hummocky grounds, sorted and unsorted circles or polygons, with or without a raised center. Their sizes vary from decametric to decimetric. They tend to develop on flat surfaces but can also form on slopes through stretching and solifluction. They are strictly related with seasonal frost activity, with or without permafrost and the presence of moisture. Many hypotheses have been proposed as to how they develop – the commonest one is load casting associated with liquefaction (Cegła & Dżułinski, 1970; Dżułinski & Walton,

1963; Eisseman, L., 1981; French, 1986; Kuenen,1958; Murton & French, 1993; Shilts, 1973; Vandenberghe, 1992). Cryogenic deformations are different in processes from thermokarst involutions resulting of downsagging by liquefaction (Murton & French, 1993) or load casting due to seismic activity induced by deglaciation (Van Vliet-Lanoë et al., 2004). Also thermal and hydraulic convection has been advocated (Mackay, 1980; Hallet & Prestrud, 1988; Hallet & Waddington, 1992), but have never been demonstrated. Already, in 1932, Mortensen recognised that temperature-controlled convection in soil was too weak to induce soil circulation. Moreover, most of these authors only focused on patterned ground developed on permafrost (Mackay, 1980; Zoltai, 1973), despite them being active and frequently found in non-permafrost terrains (large polygons excepted). The oldest hypothesis for ground patterning is deformations resulting from differential frost heave (Mortensen, 1932; Pissart, 1976; Sharp, 1942; Solovyev, 1973; Washburn, 1956). Patterning processes were visualised both in the field and by microfabric analysis (Van Vliet-Lanoë, 1985, 1988b-c, 1998), confirming the laboratory experimentation of Pissart (1982) before being further demonstrated by modeling (Peterson & Krantz, 2003, 2008). Differential frost heave is now considered as the main trigger for ground patterning, with or without permafrost, sometimes with the help of thermal or desiccation cracking and, always forgotten, of pedogenesis.

Before discussing the impact of climate change on ground patterning, we shall first have a brief look at the different contributing mechanisms to their formation: frost and permafrost, differential frost heave, ground drainage and other cryogenic processes, soil forming and biogenic processes. We shall also try to homogenise and simplify the vocabulary, and show the different typologies revealing the same process of formation.

1. The Frost Rules

1.1. Ground Freezing

Ice can nucleate at the surface of a wet soil during quick cooling and form pipekrake or needle ice, that lifts stones at the surface of the ground, or at depth, in desiccation microfissures or nets (retraction), segregates ice lenses or reticulates ice blades.

Seasonal frost may develop from a few centimetres to 2–3 meters depending on the region, its altitude, its latitude, the grain size of the ground and the climate. Usually, it reaches a deeper horizon with rising altitude or latitude. Frost penetrates from the ground surface; thaw proceeds from the surface and also the base of the frozen layer. Its cyclic activity may be daily, as in tropical mountains, to yearly, as in polar latitudes.

Permafrost is a part of the subsoil which remains frozen (T°<0°C) during at least two successive years. The upper part of the soil melts each summer; it is the active layer. The depth of the active layer is controlled by climate and moisture content, being deeper in drained terrains, and shallower in waterlogged or poorly drained soils. The upper part of the permanently frozen ground is called the permafrost-table. It is often rich in ice accumulation for the first 10 m of the soil in interglacial condition, and much more during glacial events. Permafrost records the thermal deficit induced by climate, especially in summer on a long time-span (1000 years). The permafrost is considered as sporadic when it only occurs in peat bogs, depressions or on north-facing slopes. It covers less than 30% of the landscape. The Mean Annual Air Temperature (MAAT) is >-3°C and the Mean Annual Soil Temperature (MAST) is close to 0°C. The permafrost reaches a thickness of about 20–30 m. Discontinuous permafrost covers less than 80% of the landscape. The MAAT mostly ranges from -3°C to -7°C and the MAST from 0°C to –3°C, controlled by the snow cover. The depth of the lower boundary of the permafrost ranges from 20 to about 60 m. Continuous permafrost covers more than 80% of the topography. The MAAT is <-7°C and the MAST is <-3°C, varying with the dynamic of the snow cover. Every year, by late autumn, the active layer or the seasonally frozen ground starts freezing both from the soil surface and from the permafrost-table. The following spring, the soil drains progressively from the onset of thaw leading to a shrinkage pattern expression.

1.2. Ice lensing

The process responsible for the development of frost heave is the segregation of ice in the form of lenses or veins (Penner & Goodrich; 1981; Penner 1986; Rempel, 2007; Taber, 1929; Tsytovich, 1975). This ice nucleates in fissures formed by desiccation, which opens parallel to, and below, the freezing front as frost penetrates the moist soil material. Water migrates in the direction of the higher suction, towards the cold soil–air interface: thermal suction is induced by a thermal gradient and cryosuction is induced by the formation of ice, causing water to move towards the freezing front. Once a fissure is transformed by ice lensing, it is commonly reoccupied by ice during further freeze–thaw cycles and enhanced by mechanical stresses related to the ice accumulation (Van Vliet-Lanoë, 1985, 1988b). Since it is related to successive freeze–thaw cycles in soils, porosity is progressively enhanced but becomes stable after 4–5 cycles. Segregation ice is responsible for various cryogenic structures formed by ice lensing, including foliated and platy structures, and angular and prismatic structures formed by reticulate segregation ice (Tsytovich, 1975). A granular structure originates by cryogenic stresses associated with cryoturbation and solifluction, especially in clay-rich materials (Van Vliet-Lanoë, 1976, 1985). Most of the cryogenic (micro) fabrics are thus linked: to shrinkage related with desiccation induced by low vapour tension and crystallization; to frost heave pressure; to gravitational movements during thaw consolidation (sliding or setting).

1.3. Frost Heave and Frost Susceptibility

The ice segregation potential or intensity of development of segregation ice is clearly linked to the ability of water to migrate slowly towards the freezing front or to the growing ice lenses, below the freezing front, even at temperatures below 0°C. In turn, it is directly responsible for the rate of frost heave of the soil (Black & Miller, 1985; Gilpin, 1980; Glen, 1974; Miller, 1972). Water retention and supply are the most important factors (Kaplar, 1974). It is thus clearly linked to the grain size of the sediment and its porosity. During the process of ice segregation, the water content, in the form of segregated ice, increases considerably to reach between 40 and 80% of the soil volume. In a thaw, the soil is fragmented in a cryogenic fabric and oversaturated in water: without lateral drainage, it may sometimes liquefy.

A sediment, or soil, is called frost susceptible if it is able to develop many ice lenses and thus able to heave during progressive and slow frost. Fine-grained sediments, from silts to fine sands, are favourable for water migration and thus important for ice segregation. The most frost-susceptible soils are fine grained, heterogeneous in particle size distribution with at least 5% colloidal content (Kaplar, 1974) especially when the water content is at field capacity (maximum of free water retention), and with a pF value around 2.5 in mineral soils or close to 1 in organic soils (Mériaux, 1979). Tills, loessic materials and spodic soil horizons (podsols), rich in amorphous clays, are very frost susceptible.

If the water supply is small (arid climate or coarse grain size), the permeability is low (as in clays) or the possibility for water adsorbed at the surface of the minerals (as in sands) to migrate is limited, no shrinking and almost no segregation of ice will take place and the soil will not heave during freezing. They are considered non-frost susceptible.

1.4. Frost Susceptibility Gradient

Minute differences in water retention or permeability can lead to differential frost heave (Sharp, 1942; Washburn, 1962; Dylikowa, 1964; Van Vliet-Lanoë, 1985), shown in profiles by deformations or cryoturbations, noticeable on flat land surfaces as well as on slopes. The finer material will develop more ice segregation, due to water migration to the freezing front and this delays frost compared to the coarser material. Differential frost heave results from the juxtaposition or the superposition of material with different frost susceptibilities.

Frost susceptibility contrasts and gradients control the geometry of the deformations. The gradient will be positive if the surface material is more frost susceptible than the underlying sediment (via differences in clay, silt or organic content). The gradient is clearly negative in the case of an illuvial subsurface horizon or of a superficial accumulation of frost-jacked stones. In the first case, the contrast of frost susceptibility will be weak. In contrast, it will be major between the stony superficial layer (Figure 1) which behaves as a non-frost-susceptible material on the underlying frost-susceptible material.

Figure 1. Sorted circle in section, Banks Island, NWT Canada (photo A. Pissart) and from above, (Brandalpynten, NW Svalbard) showing differential frost heave with a negative gradient of frost susceptibility.

Particle translocation during ground melting may modify frost susceptibility with time (Van Vliet-Lanoë, 1988a). Fine particles produced by frost shattering (see 1.5) and translocated vertically, and/or laterally, result in accumulations of silt and coarse clay at a shallow depth. Particle translocation is strongly controlled by the suction intensity resulting from the thermal inversion during the melting season: the coldest horizon is located at depth. Because of decalcification, particle translocation and changes in porosity, horizons with different water retention or frost susceptibility develop with pedogenesis. Eluviation proceeds deeper with the ageing of the surface inactivating very old patterned ground

Biological activity is mostly confined to the warmest season, and bioturbation is seasonally active and alternates with ice lensing. It integrates humic and other substances into the soil matrix and increases soil water retention and thus, frost susceptibility. Because of humus incorporation, and changes in porosity, soil horizons with different water retention or frost susceptibility develop with biological activity. Patterned ground, such as mud boils can develop in seasonal alternation with biological activity and weathering. With lower altitude or latitude, biological activity and weathering prevail on frost-induced processes (Figure 2) for the formation of soil mounds.

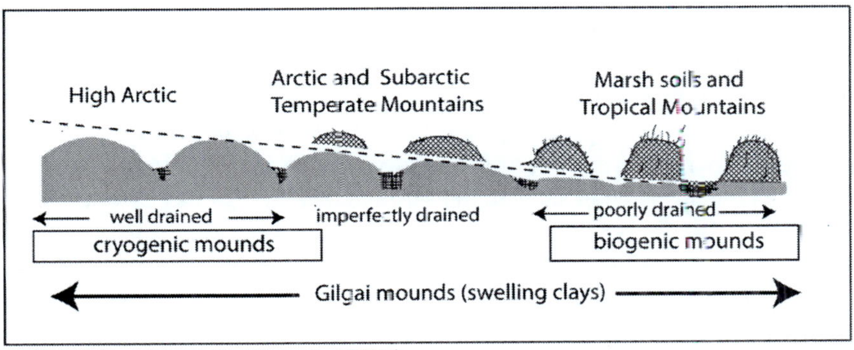

(Modified from Van Vliet-Lanoë, 1988c).

Figure 2. Relation between frost mound or hummock, the biological activity and the mineralogy of the clay fraction.

In Alpine mountains, or at lower latitude, cryogenics is completed by biogenic mulching, mostly by ants or termites. In savanna or chaparral, biogenic mounds follow the same repartition – self-organised and related with moisture. A latitudinal/altitudinal gradation exists from the pole to the equator, from purely cryogenic forms to purely biogenic ones. It can be complexified if the mineralogy of the clays is smectite (gilgai: converging morphologies; Van Vliet-Lanoë, 1991). Also mima mounds in the American meadow seem to be mostly controlled by the biological activity (Washburn, 1988).

1.5. Other Contributing Processes

Snow depth is a major factor as it influences the vegetation cover (Arsenault & Payette, 1992; Egginton, 1981; Lull & Rushmore, 1961; Rissing & Thorn, 1985; Tedrow, 1977; Walker et al., 2001; Zoltai, 1973). By insulating the soil surface when dry, it protects the ground and the soil fauna from important freezing and also from thermal cracking (Washburn, 1979; Payette et al., 1986). For this reason, snow depth, as with peat and moss, limits the surface expression of differential frost heave and their disparition favours it (Egginton, 1981; Kade & Walker, 2008; Rissing & Thorn, 1985; Sigafoos, 1951). In contrast, in a dry environment, the snow patches kept by the vegetation provide moisture in springtime for the *secondary frost heave* (Black, 1995; Fowler, 1989; Parmurzina, 1980; Mackay, 1980): melt water migrates to the still frozen horizon, enhancing the initial frost heave (Hoekstra, 1966; Williams & Wood, 1982). This is very sensitive in hummocky terrain or in snow patches (Mackay, 1980; Woo et al., 1982).

Desiccation cracks develop during thawing in more or less frost-susceptible material, in connection with drainage, creating small polygonal forms up to 1 m in diameter (Figure 3a-b). Desiccation cracking results in decametric to metric nets during the summer thaw and drainage when soil moisture content reaches air penetration potential from initial saturation (Van Vliet-Lanoë, et al., 1984). In fine grounds (e.g. lacustrine or tidal mud), large decametric polygons may form outside the range of permafrost (Butzer, 1971). Frost fissures or desiccation fissures are the most common features and at the starting point for most of the periglacial deformations.

Thermal cracking creates polygonal decimetric to decametric networks (Lachenbruch, 1966; MacKay, 1974; Schunke & Zoltai, 1988; Washburn, 1979). Thermal contraction also needs abrupt cooling of more than 10°C within a few hours to develop (Lachenbruch, 1966; Mackay, 1974). To induce in sediment an efficient thermal cracking, high ice content is basically necessary, as ice displays the largest coefficient of thermal retraction compared to other natural rocks. Polygonal wedges develop easily in frozen peat, in a highly frost-susceptible material (Figure 3c), as it will in sands or gravelly sands, and essentially in waterlogged conditions. Ice veins and initial thermal nets develop during low water in alluvial plains or beaches. Thermal cracking may occur in non-permafrost grounds (Washburn et al., 1963; Svenson, 1977).

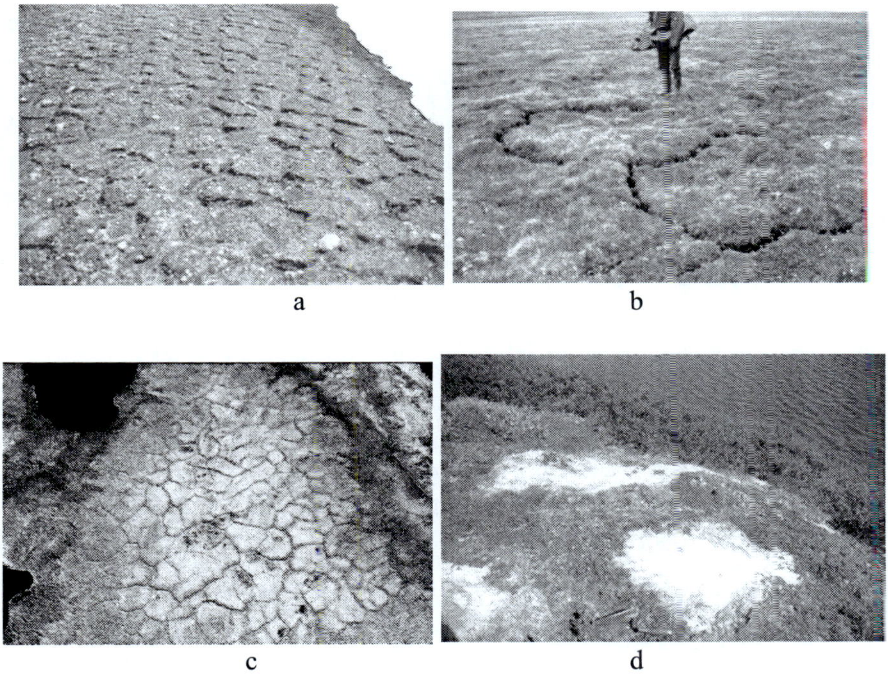

Figure 3. Polygons a) desiccation polygon (decimetric) in loess Advent, Svalbard) b) (glaci-marine silt on permafrost, Advent, Svalbard), c) very large thermal polygons on ice wedges (35m ø, on permafrost), Tuktuyaktuk, NWT, Canada (Modis, NASA), d) crusting and punctual liquefaction of mudboil (Lake Boniface, N.Ungava).

Liquefaction is rare, even in "mud boils" (Shilts, 1978, see § 3.1.2) with unstable cryogenic fabric (Van Vliet-Lanoë, 1988b) and is generally related to high interstitial water pressure resulting from downslope hydrostatic pressure at thaw (Figure 3d), and loading by deposition of fresh sediment or animal trampling. In fact, most figures miming liquefaction are soil crusting formed by soil surface fabric collapse under rainfall as in other climatic regions. Load casts are uncommon, except when related to unconsolidated, water-saturated, freshly deposited sediment. Vibration associated with earthquakes, trampling or traffic can accidentally destroy the cryogenic fabric and the soil then exceeds the liquefaction limit.

Frost shattering is the activity of frost on rock and rock fragments. The mechanism for frost shattering includes frost wedging (Washburn, 1989), ice segregation (Murton et al., 2006), hydraulic fracturation (Everett & Haynes, 1965; Powers, 1965) and rock fatigue. It produces clasts which reduce progressively in size with successive frost cycles to reach a minimal value (Lautridou, 1985). It is active on buried rock (Figure 4) or even when the clast is incorporated into the ground matrix (Van Vliet-Lanoë, 1998c). It works in competition with stone frost jacking and soil-forming processes such as particle translocation (Locke, 1986; Frenot et al., 1995; Jacobsen, 1989) or decalcification. Stone frost jacking recorded by the vertical orientation of stones is common and results directly from the vertical movement of stones by ground frost heave and thaw settlement (Beskow, 1935; Corte, 1966; Pissart, 1969). Its favoured by elongated clasts morphologies. It may be helped by soil weakness as polygonal cracks (thermal or desiccation). This mechanism can lead, with particle leaching and frost shattering, to the creation of a negative contrast in frost susceptibility by the formation of a coarse, non-frost-susceptible surface layer resting on the accumulated residual ground. Stones have a high thermal conductivity as they are usually poor in the retention of water. They can thus locally modify the soil thermal and hydraulic properties both at the macroscopic and microscopic scale.

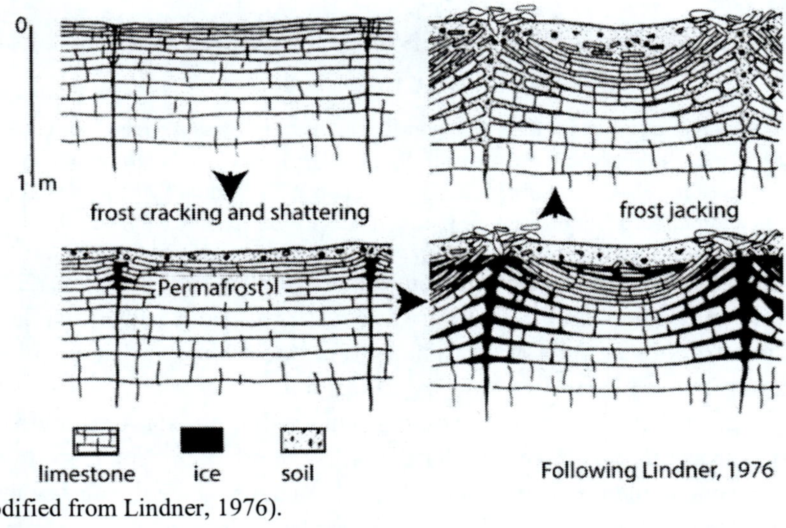

(Modified from Lindner, 1976).

Figure 4. Impact of frost shattering and thermal cracking on sorted polygon on shallow limestone bed-rock.

a b

Figure 5. Solifluction deformations. a) patterned ground on raised beach, located on permafrost and stretched by solifluction (Svalbard) and b) soliflucted patterned ground: shales on dolostone (La Mortice, 3500 m Southern French Alps).

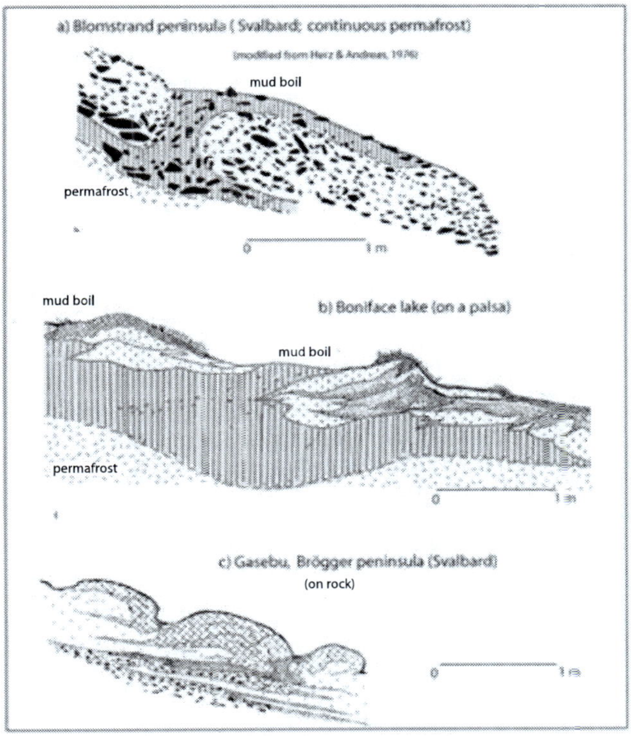

Figure 6. Solifucted forms. a) Sorted circles (Blomstrand peninsula, NW Svalbard) b) Circles on a silty palsa (Boniface lake, N. Québec) c) cushion-like solifluction of hummocks on permafrost (Gasebu, NW Svalbard).

Periglacial solifluction is mass movement controlled mostly by frost creep which results from combined thaw settlement and sliding on slopes under gravity; it is directly related to the content in segregation ice and to water supply during the thaw (Benedict, 1976; Coutard et al., 1989; Harris, et al. 2003; MacRoberts, 1978). Frost creep has no specific climatic association as only frost is needed.

Deformations, or the position of stones, are related to ice content and moisture supplied during the thaw: stones lying flat with long axes oriented in the direction of the slope indicate excess moisture by snow melt; vertical stones indicate dominant frost heave and stone jacking. Solifluction is controlled both by slope and water supply, but it is increased by differential frost heave due to particle lateral translocation and a consequent increase in frost susceptibility. Solifluction may activate from a very low angle slope (>3°), alternating in season with cryoturbation and ground patterning (Figure 5 a-b and 6).

2. THE ROLE OF DRAINAGE

The role of drainage is of major importance as it will enable the development of ice lensing by supplying moisture from the depths, from melting snow at thaw. Water will be available as:

1) Interstitial (ground saturation by the water-table)
2) Water film within the capillary fringe of the water-table, corresponding to air entrance into the sediment or pF=1
3) Above the water-table, moisture is available both by suction and vapour phase, corresponding to the field capacity or pF=2.5. Water films persist, allowing fair development of the ice lenses and thus the differential frost heave.
4) Thin water films adsorbed at the surface of colloids and minerals, more or less available for feeding the growing ice crystals.
5) A vapour phase that can migrate to the coldest part of the ground and crystalize on the largest existing pores and ice crystals (curvature effect: pore and crystal).

The water exiting within the minerals is rarely used in the ice crystallization process.

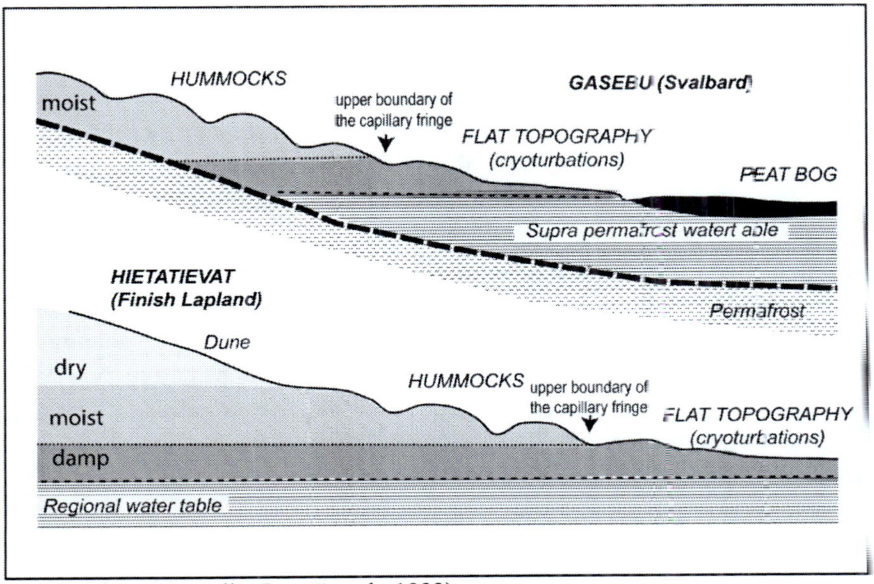

(modified from Van Vliet-Lanoë et al., 1992).

Figure 7. Relation between the topography and the drain-age in autumn (onset of the frost season).

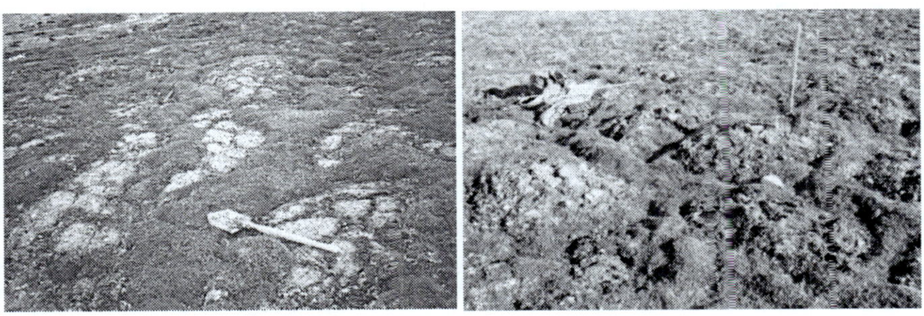

Figure 8. Unsorted circles on a former lacustrine depression, on permafrost, claysish silt (foot of the Zepplinfjellet, NW Svalbard).; a) poorly drained, b) drained (1m higher in altitude).

When the soil is dry – the capillary films on the grains being disconnected – ice only forms sparse crystals in the intergranular porosity: the soil is not cemented when frozen.

When the ground is at field capacity, a fair microrelief may develop due to retraction during the drainage of water in excess after thaw, and due to the best conditions for developing differential frost heave. This is the zone of raised forms, hummocks and other frost mounds (Figure 6c, 7 and 8b).

When the ground is in the capillary fringe of the water-table, some microrelief develops in connection fissures or the presence of stone, which induce a positive deformation in the frost penetration front. At depth, the differential frost heave is efficient to produce cryoturbations.

When the ground is saturated, or the soil is very porous (as in sands), it freezes in-situ with virtually no migration, thus with a limited heaving related to the change in volume of the water (10%). These wet grounds freeze as a mass, or a block, and behave as non-frost susceptible, explaining flat morphologies for patterned ground. Soils and patterned ground in depressions remain flat with no microrelief (Figure 7 and 8a). Nevertheless, at depth, the differential frost heave may develop when the frost penetration slows down, and cryoturbations develop.

The permafrost-table creates a summer barrier for surface waters. It constitutes a variable lower limit for the water-table and for translocation. The geometry of the permafrost-table is directly linked with soil surface properties and vegetation cover which may fix snow accumulation (see § 3.1). However, ice accumulation within the permafrost at depth can modify the topography and soil drainage, leading to gleying or solifluction, inconsistent with the post-permafrost topography. The occurrence of permafrost has an impact on the hydraulic functioning of the subsoil. Fluids migrate from the subsoil to the coldest part of this horizon, at a depth of about 10–20 m, which is that of penetration of the annual thermal fluctuations, increasing the ice content of the upper horizon of the permafrost.

This system is valid as well on permafrost as in seasonaly frozen grounds (Figure 7). In a similar sedimentary and pedological environment, following the position on the slope, cryoturbations develop by differential frost heave in relation with a poorly drained to waterlogged position, although raised forms such as hummocks and circles develop in imperfectly to moderately well-drained sites, with or without permafrost.

3. Patterned Grounds

Patterned grounds cover various morphological forms as decametric to metric polygons, stone circles or mud boils, frost mounds or hummocks, ground stripes and steps on slopes. Initiation of patterned ground is mostly related to a net (regular field) or random (coalescing, self-organised field). They may be sorted or not (Figure 10).

3.1. Unsorted Polygons, Hummocks and Circles

3.1.1. Polygonal patterns

Polygonal patterned grounds, without specific micro-relief, are usually related to thermal cracking from metric to decametric size (Figure 3). They form with simple cracking, ice veins or ice wedges, sand or gravel wedges. For ice veins and ice, they are strictly linked to permafrost, with yearly accumulation at depth of ice in the crack, with the permanently frozen ground.

Metric to decimetric forms are related with desiccation during melting, and are enhanced by frost desiccation in the autumn and by secondary heave in the spring. They may be flat in a poorly drained position or evolve into frost mounds or hummocks in drained one (see § 2.1.3; Figure 7). The geometry of the deformation is similar to those of gilgaï (Paton, 1974; Van Vliet-Lanoë, 1988b, 1988c). They are easily reproduced by experiment (Pissart, 1966). The dominant pattern is constrained by the lithology of the unconsolidated substratum (Brown & Gangloff, 1980; Van Vliet-Lanoë, 1988a Figure 10).

3.1.2. Circles and Mud Boils

Circles and mud boils correspond to random earth boils at the ground surface, isolated (Figure 10) or in large fields, and in non-gravelly grounds. Circles may develop as self-organised nets, and merging as coalescing, more or less circular, patches, resulting from the emersion out of the subsurface material in a negative frost-susceptibility gradient. These forms can extrude from various depths, on permafrost as on shallow basement or even in the vicinity of a shrub allowing the accumulation of snow (Figure 11a).

Frost susceptibility contrasts, and gradients, control the geometry of the deformations. Following the gradient of susceptibility, coalescing, self-organised circles can be obtained with a negative gradient. For random configuration, pattern initiation can theoretically begin at any depth of freezing. A shallow basement may be sufficient to keep a shallow water-table and permit the expression of random, self-organised mud-boil fields (Figure 11b). They may protrude irregularly from below the gravels of a garden on permafrost, as on the Kenai Peninsula (Homer, Alaska, Figure 11c).

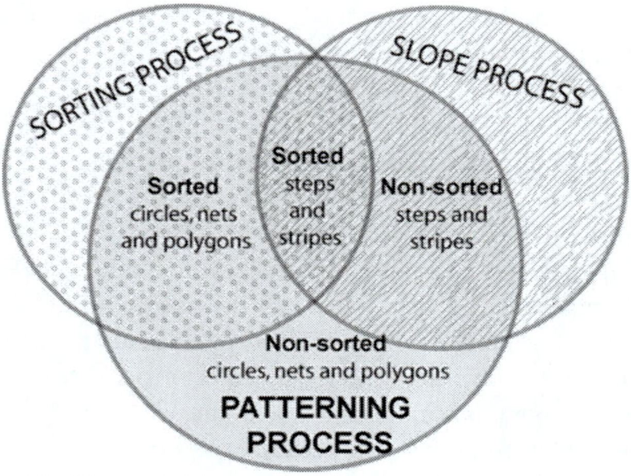

(Modified from Washburn, 1979).

Figure 9. Venn diagram showing the interaction of ground patterning, sorting and slope process.

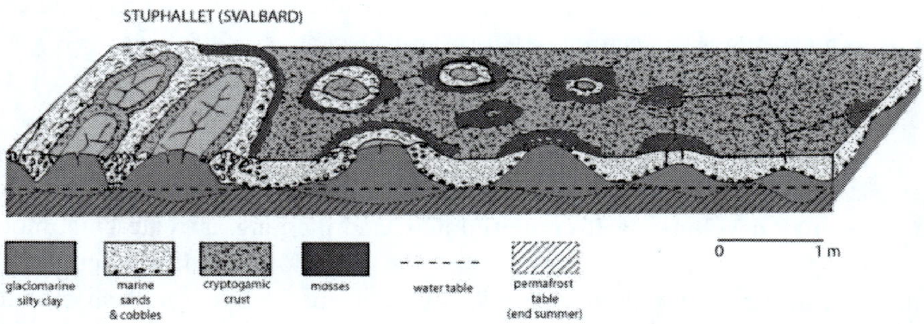

Figure 10. Example of the relation between the development of patterned ground and the geometry of the substratum (raised beach, Stuphallet; NW Svalbard) (completed from Van Vliet-Lanoë, 1988a).

Circles usually form by differential frost heave on superposed sediments with various frost susceptibilities or capacity to heave under the action of frost, but with a distinct negative frost-susceptibility gradient. It is the typical condition for a cryoturbated ground in section.

3.1.3. Cryoturbations

Cryoturbations are linked to ground patterning and reflect in section the downward deformation in imperfectly drained to waterlogged soils, in relation to the rigid comportment of the surface layer at the onset of frost (Figure 11); in drained sites, superficial differences of frost heave are expressed upward and the geometry of residual stresses leads to the creation of hummocks or frost mounds decimetric to metric in size. Contiguous forms are also common. Polygons may just form alone, initiated by frost desiccation or by thermal contraction, in a positive or negative frost-susceptibility gradient. They are anchored to a deeper layer with a negative frost-susceptibility gradient.

Contiguous circles may develop as well-organised nets, initiated by frost desiccation or by thermal contraction. Emersion of the subsurface material in more or less circular patches is usually constrained in space by the nodes in the cracking pattern, usually a little more drained, and thus freezing at first. The extrusion usually begins in early spring due to melt-water refreezing (regelation) at a shallow depth in the ground. Once formed, the extrusion develops from the onset of the next frost seasons. Cracking guided nets (thermal or desiccation) are perennially stable. In vegetated landscape (tundra or meadow) a rim of vegetation may form, similar to stone circles, and broken in the case of a coalescing of two mud boils.

a

Figure 11. (Continued)

Figure 11. Unsorted circles. a) mud boil at the place of winter' snow accumulation; no permafrost; b) mud boil on a very shallow bedrock (a & b : Boniface lake; discontinuous permafrost, Northern Québec). Young mud boils (silts) in a garden , on discontinuous permafrost (Homer, Alaska).

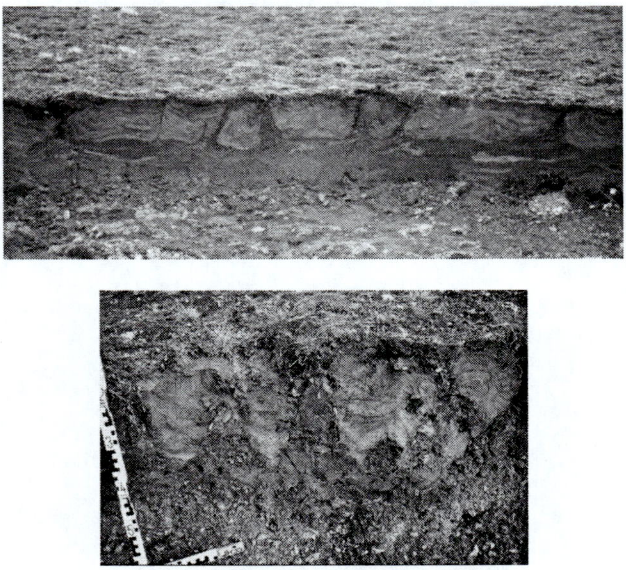

Figure 12. a) Flat topography with cryoturbation, at left, positive gradient of frost susceptibility in Aeolian sands; b) negative gradient: the upward moving organic silt is piercing the surface forming locally a mudboil. Poorly drained site on continuous permafrost; Gåsebu, NW Svalbard.

3.1.4. Role of the drainage

Both contiguous and coalescing (self-organised) forms have a surface that remains flat in poorly drained ground and sometimes with a decimetric secondary polygonal net of desiccation at its surface. In a drained situation, they evolve into raised forms or hummocks (earth hummocks), sometimes lifting up the thin peat cover (peaty hummocks, *pounus*).

Tarnocaï and Zoltaï (1978) have shown that for raised circles in silts, the distance between the hummocks is less than the diameter of the polygonal crack. Mud boils are circles with a very unstable surface grain size, and are highly susceptible to soil crusting under splash and to liquefaction (Shilts, 1978). On slopes, they may form lines of stretched mud boils, somewhat rimmed by lichen and mosses or grassy vegetation. If the slope is above 5°, step-like terraces may develop.

3.1.5. Hummocks or Frost Mounds

Frost-formed mounds, or hummocks, typically exhibit some variations across temperate mountains, sub-Arctic and Arctic environments (Figure 13a). In sub-Arctic earth hummocks, such as the Alpine one (Crampton, 1977; Mackay, 1980; Schunke & Zoltai, 1988; Van Vliet-Lanoë, 1983, 1988b), mound development requires imperfect to fair drainage, frost-susceptible material and a climate that allows irregular snow distribution. Vegetation development seems to be an important factor in enhancing growth, once formed, due to the snow accumulation between the mounds (Figure 11a). This allows the development of secondary frost heave in springtime, sometimes "exploding" the mounds as in the Alps or in Iceland.

These forms are as frequent in sub-Arctic regions as they are in temperate mountains (meadow zones; Figure 13b) outside the range of permafrost. In sections, they may show various morphologies, from a simple form to an injected form, or with pseudo-convective morphologies (Van Vliet-Lanoë et al., 1993; Van Vliet-Lanoë et al., 1998; Figure 14).

On slopes, they are elongated along the direction of the slope by subsurface frost creep as "sliding cushions" (Figure 6c), sometimes aligned in step-like rows.

Thufur (Shunke, 1977; Van Vliet-Lanoë et al., 1998; Figure 14c) are usually raised polygons that develop in highly frost-susceptible volcanic loesss of Iceland. They may appear 2–3 years after ground perturbation. Due to their high ice content in winter, they are often incised by a spring decimetric desiccation net. In the Massif Central (France), they may form at altitude (ca 1000 m) helped by the activity of the soil fauna (organic matter integration).

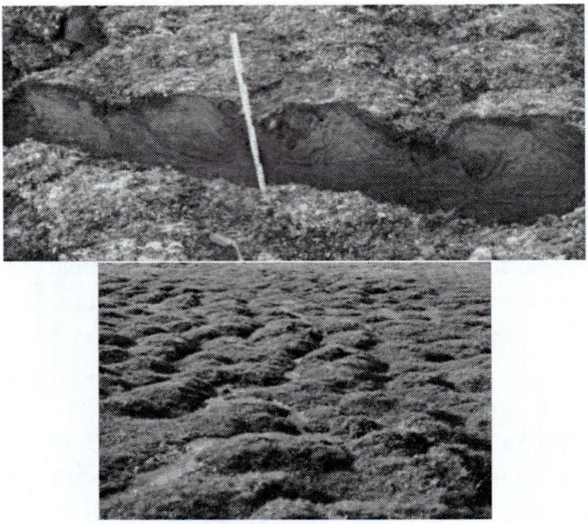

Figure 13. Sandy hummocks or raised center circles on well drained site oncontinuous permafrost Gåsebu, NW Svalbard; b) silty earth hummocks at the Valonnet, 2500m, Southern French Alps (no permafrost).

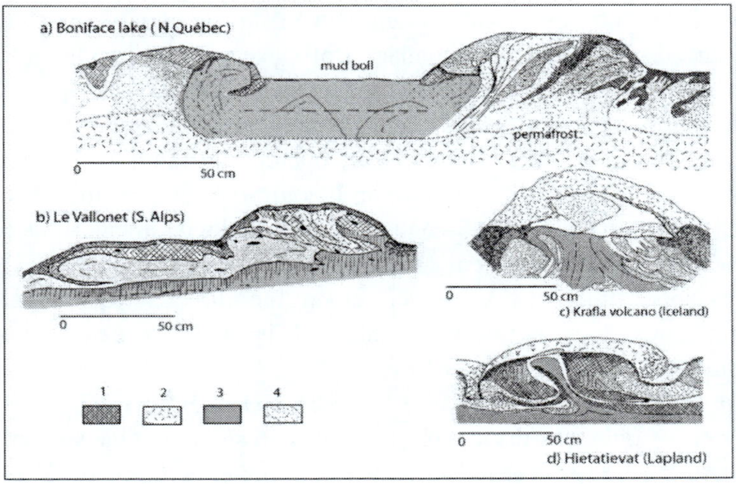

Figure 14. Various form of hummocky grounds and non sorted circles. a) a low center circle on thin and recent permafrost; b) earth hummocks (col du Valonnet; 2500 m, Southern French Alps, no permafrost); c) "thufur" (no permafrost, 400m, Krafla volcano , Northern Iceland); d) sandy hummock (no permafrost, 200 m, Hietatievat, Finish Lapland). Notice in most figures, some limited lateral deformation by solifluction. (legend: 1- organic; 2-frozen, 3- silty loam, 4- sand).

Pounus are frost mounds, often connected with vegetation growth capability with a thin peat cover (Van Vliet-Lanoë & Sëppala, 2002). They are random superficial forms with variable control by frost heave, and are independent from the presence or absence of permafrost. They are related to a dynamic snow cover with open birch stands or tundra.

Most of the deformations observed in section, in the Arctic as well as in temperate mountains, reflect alternating periods of intense frost activity (available moisture) and periods of pedogenesis (less-intense frost activity). Transects in sectors affected by unsorted circles or frost mounds show that the spatial variability is closely linked to small drainage change, change in vegetation and lateral change in frost susceptibility or geometry of the different strata of the ground.

Thufur-like constructed forms, also with a random repartition, can develop by sediment fixation (aeolian, niveo-aeolian, fluvio-glacial), by humidity and/or vegetation (Van Vliet-Lanoë, 1988) from Arctic to milder regions.

3.2. Sorted Polygons and Circles

In "sorted" morphologies, the processes are identical as for unsorted, contiguous and guided by a polygonal pattern, or self-organised and coalescing, but the less-susceptible surface layer must be stony to gravelly. These forms are much more spectacular compared to non-sorted.

3.2.1. The Non-Frost Susceptible Surface Layer

The gravelly surface layer can be produced: (i) syn-sedimentary by layer superposition, (ii) by frost shattering of larger, more scattered bocks (e.g. ablation till), (iii) by frost jacking of stone from a diamicton or a tidal mud, and (iv) by pedological processes, as particle leaching and decalcification, accumulating fine particles at depth (B silt, B textural) (Van Vliet-Lanoë, 1988a; Etzelmüller & Sollid, 1991). These processes give rise to very negative gradients of susceptibility, thus to high frost-susceptibility contrast. They develop on any lithology, from sandstone to shales or granites. They are spectacular on highly frost-susceptible lithologies. This is particularly well developed on dolomitic limestone that produces a fine residual matrix (dolomitic silts and residual clays) both by frost shattering and dissolution of the calcite cement. In many places in the Arctic and sub-Arctic, they develop on raised beaches, uplifted by glacio-isostatic rebound. In Alpine mountains, it is mostly the case of frost-shattered shales, due to the higher frequency of freeze–thaw cycles (Figure 15d).

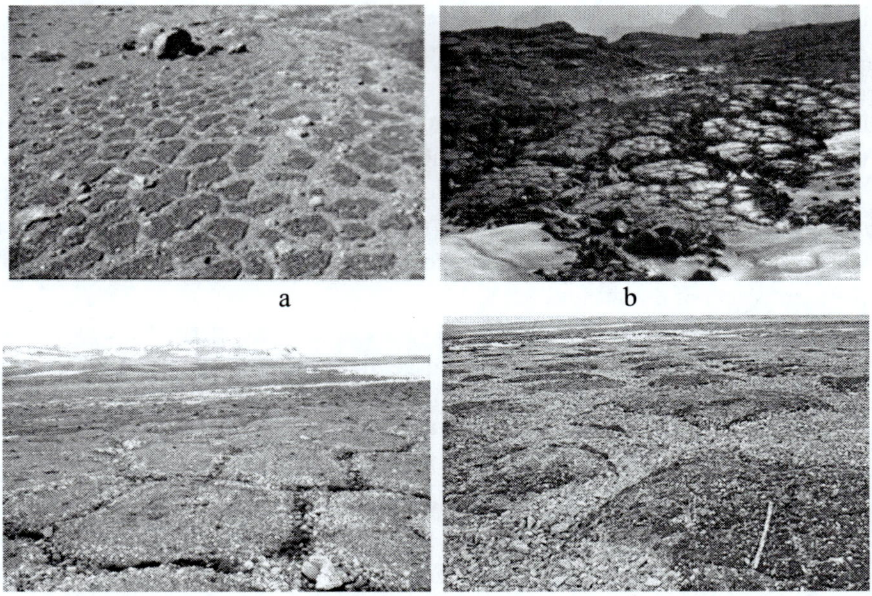

a b c d

Figure 15. Sorted polygonal patterns. a) small forms, southern Iceland; b) non-sorted and sorted polygons on shallow bedrock, La Mortice (southern French Alps) c) low center sorted polygon on permafrost (Kwadehuk peninsula, NW Svalbard), d) high center sorted polygon on permafrost (c & d: Kwadehuk peninsula, NW Svalbard).

3.2.2. The Stone Circle

The formation of the stone circle results from the lateral pulling of the surface gravel, especially when the central mud boil is heaved by frost. A reactivation may be visualised by the upturning of the stones, lacking common lichen crusting. Residual pits may survive between the stone circles. These forms correspond to the self-organised or coalescing forms.

With small polygons, frost-jacked gravels usually move to the frost crack by pipekrake uplifting or by frost creep (Pissart, 1966). On a metric polygonal net, it is the frost jacking of the frost-shattered basement that activates: the stones are jacked upwards thanks to the frost susceptibility of the surface layer and then accumulate; pits may form at the intersection of the polygons. This pattern corresponds to the "festoon-like" cryoturbation in section.

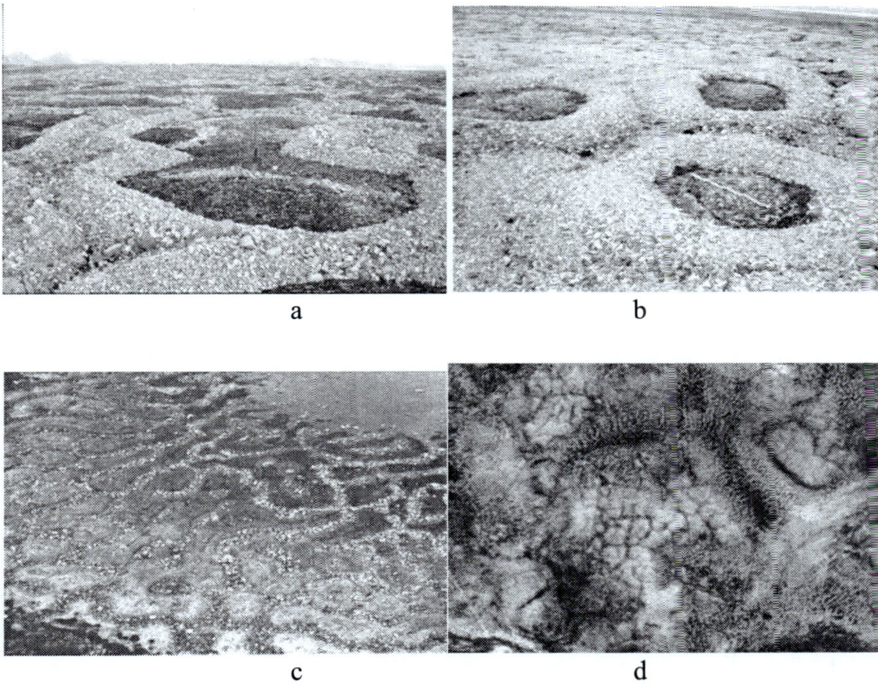

Figure 16. a) Coalescing or self-organised sorted circle and b) contiguous sorted circle on Kwadehuk peninsula, NW Svalbard. d) patterned ground development on the bank of a subarctic lake (Northern Québec, photo J. C. Dionne); sorted circles in lacustrine depression and polygons on the drained surface, Kwadehuk peninsula (NW Svalbard).

3.2.3. The Frost-Susceptible Deep Layer

The subsurface layer produced by particle leaching or translocation creates a highly frost-susceptible layer at depth that can more or less regularly pierce the surface in the form of circles (Van Vliet-Lanoë, 1988a) or of "plug circles" (Washburn, 1997). Injections usually occur in nodal position (Van Vliet-Lanoë, 1988a) or as a vertical blade of a crack net (Washburn, 1997). Cylindric injections are described as "plug" by Washburn (1997) at Cornwallis Island (NWT, Canada); they may follow a thermal crack. The injections are generally marked by vertical clasts. Similar features occur in Pleistocene raised beach as at the Crotoy (Picardy, France), related with the major permafrost extent in France, ca.250 000 yrs. (MIS 8; Figure 17 b).

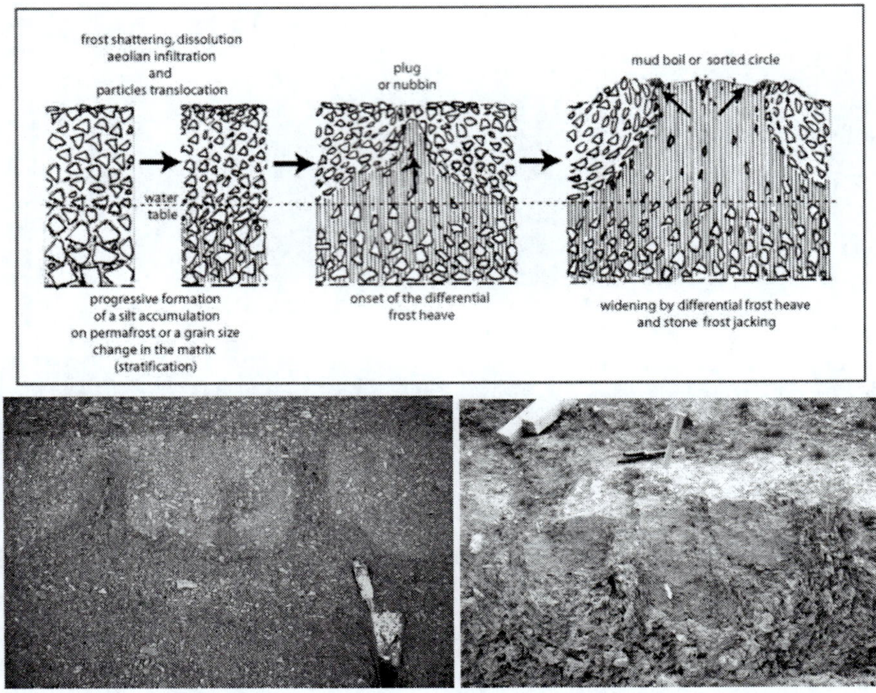

Figure 17. a) Modality of the formation of an illuvial B silt horizon at Kwadehuk peninsula (Svalbard, modified from Van Vliet-Lanoë, 1988a);b) cryoturbated Bsiltine a raised beach of Picardy (NW France). c) festoon-like cryoturbations in central Brittany (Naizin; loess on frost-shattered shales).

Particle leaching at thaw may modify the frost susceptibility with time, limiting in surface the capability for ice lensing and "fossilising" the form. With ageing, the surface of the mud boil is leached in fine particles reducing its frost susceptibility. Such old circles developed 60 000 years ago on the Kwadehuk Peninsula, Svalbard, do not evolve more on the surface. Generally, the surface of the boil is colonised by lichens and mosses. An experiment was performed in 1964 on the Kwadehuk Peninsula (Svalbard). In 1988, not a single change had been observed, nor in 2004, despite climate change and reactivation of some of the sites. This observation is confirmed by the experiment of Kade and Walker (2005) in NWT, Canada.

3.2.4. Influence of the Drainage: Raised Centre Forms

For moisture availability, patterned ground mostly develops on shallow permafrost, as in Svalbard, or bedrock, as in the Alps, which retains a perched water-table in late summer, providing moisture for the winter frost. Any change in the annual thermal budget may modify the late-summer position of the water-table and the morphology: low-centre or raised-centre forms (Figure 18).

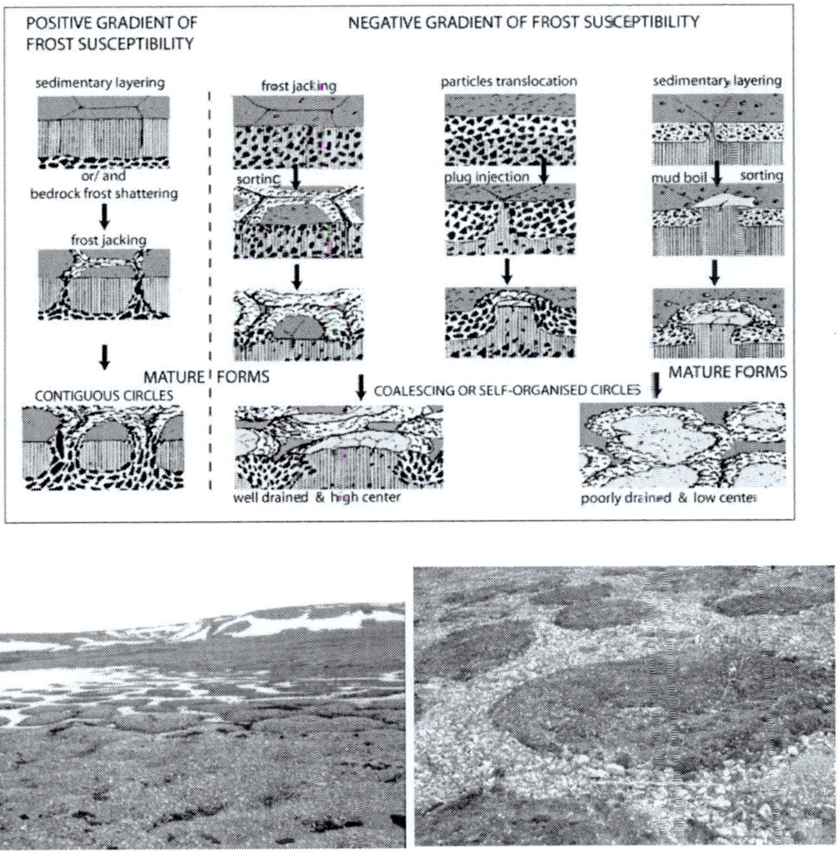

Figure 18. Sketch of the formation of sorted patterned ground (modified from Van Vliet-Lanoë, 1988a). b) and c) sorted polygones sith raised centers related with the drainage of lake (Kwadehuk peninsula, NW Svalbard).

Here too, most of the deformations observed in section, as well as in Arctic or in temperate mountains, reflect alternating periods of intense frost activity (available moisture) and periods of pedogenesis (less intense frost activity). Deformations may be quite complicated with injections mimicking "convection", like unsorted forms (Figure 14; cf. Hallet & Prestrud, 1988, Hallet & Waddington 1992).

On slopes, we may find sorted stripes in a poorly drained context, with rapid solifluction. It is the case in Alpine or Andean mountains, with a more important control by frost creep due to more frequent freeze–thaw cycles. We may also find imbricated terraces on high-latitude permafrost, especially for self-organised morphologies (Figure 6a), due to less frequent and effective frost creep.

4. Patterning and Climate Change

Ground patterning takes time, with the exception of small sorted or unsorted polygons (Ballantyne & Matthews, 1982). Major fields of patterned ground in Kongsfjorden (Svalbard) may be as old as 60 000 yrs. BP at the Kwadehuk Peninsula (Forman & Miller, 1984), although those of Brandalpynten have only ca 9,500 yrs BP and are much more dynamic. Those described at Cornwallis Island (NWT Canada) range from 4000 to 10 000 yrs. BP (Washburn, 1997).

In the Alpine mountains, their formation may be much faster due to the higher frequency of freeze–thaw cycles, but they are generally shallower (shallow frost penetration). Forms on permafrost are commonly larger and more mature that out of its limits.

For a well-expressed polygonal ground, a minimum of 1000 yrs. is required (Allard & Kasper, 2001; Mackay, 1974; Sletten et al., 2003), but at the southern boundary of the present-day permafrost they may be polygenetic and/or inherited (Payette et al., 1986; Thorarinsson, 1964; Svensson, 1973 &1977; Van Vliet-Lanoë, 2005) as is the case at higher latitude (Lewkowicz, 1994; Mackay, 1990). Nevertheless, thermal polygons are seldom in mountains, because the lack of flat surfaces susceptible to storing water and/or accumulating ice.

When the climate evolves, the main modification will be related with change in the thermal budget, impacting the freezing depth, and in precipitation, impacting the frost heave. This is very sensitive for mountain patterned ground as in Himalaya (Kühle, 1983), due to permafrost shift in altitude with climate change. Furthermore, the change in precipitation may be related with snow accumulation modality, resulting in restrictions in the vegetation cover (shrub prostration, pipekrake activity striping the ground) and also impacting the secondary frost heave in springtime. For these reasons, active patterned grounds are very sensitive to climate change. A lowering of the water-table creates hummocks and sometimes pseudo-convective forms (with or without permafrost). This is common in Nordic regions, Alpine mountains and temperate mountains (Figure 14).

Thermokarst is rarely sensitive at the ground surface, as such warming regularly lowers by less than 2 m the permafrost-table. It is only sensitive with the presence, at depth, of ice wedges or massive ice (e.g Murton & French, 1993), creating gullies or alas (thermarstic depression). A slight warming leads to the melting of the permafrost and progressive deformations with renewed pedogenesis in less continental conditions as in N. Québec, or in Ungava. With the recent warming, usually the permafrost-table becomes deeper by 50 to 100 cm, as between 1983 and 2004 in Brandalpynten and the Kwadehuk Peninsula (Svalbard, 79°N, continuous permafrost), draining the active layer in former waterlogged positions and favouring a differential frost heave (Figure 19): the circle is pulled outward and the central boil raises. The sorted circles in former drained positions inactivate. In mountain area, they are usually considered as trace permafrost conditions (King, 1986). It is not systematic but the forms are larger and more active on permafrost (sporadic to continuous).

Here, a lowering of the water-table induces a new occurrence of hummocks in previously flat and cryoturbated positions and often develops a plug-like central injection (Figure 18 & 19). Complex lowering and a secondary rise in the water-table, combined with differential frost heave and thaw consolidation produces some circles with hummocky rims pierced by secondary mud boils, as also at Boniface (Northern Québec, 57°N, discontinuous permafrost). These observations explain the plug-like deformations observed at Cornwallis Island (75°N, continuous permafrost) by Washburn (1997), also activating from 1985. Lakes may be drained, as observed at Kwadehuk in 2004, activating patterned ground in the freshly emerged bottom (Figure 18), but partly inactivating patterned-ground fields located higher in the topography. The complete disappearance of permafrost commonly leads to ground drainage and mostly to the inactivation patterned ground, shallow bedrock excepted. This drainage, especially in arid conditions, also favours the seedling'success. This is the case of regular spaced shrubs one the mud boil of inactivated patterned-ground in NW Siberia (Figure 20; Frost et al., 2013).

If the precipitation is raised with warming, the changes are limited, as the water-table is maintained in its autumn position. If snow precipitation increases, it will limit the main frost heave but may accentuate the secondary frost heave, favouring the growth of frost mounds and raised forms and sometimes, their "explosion" in springtime as in the Alps.

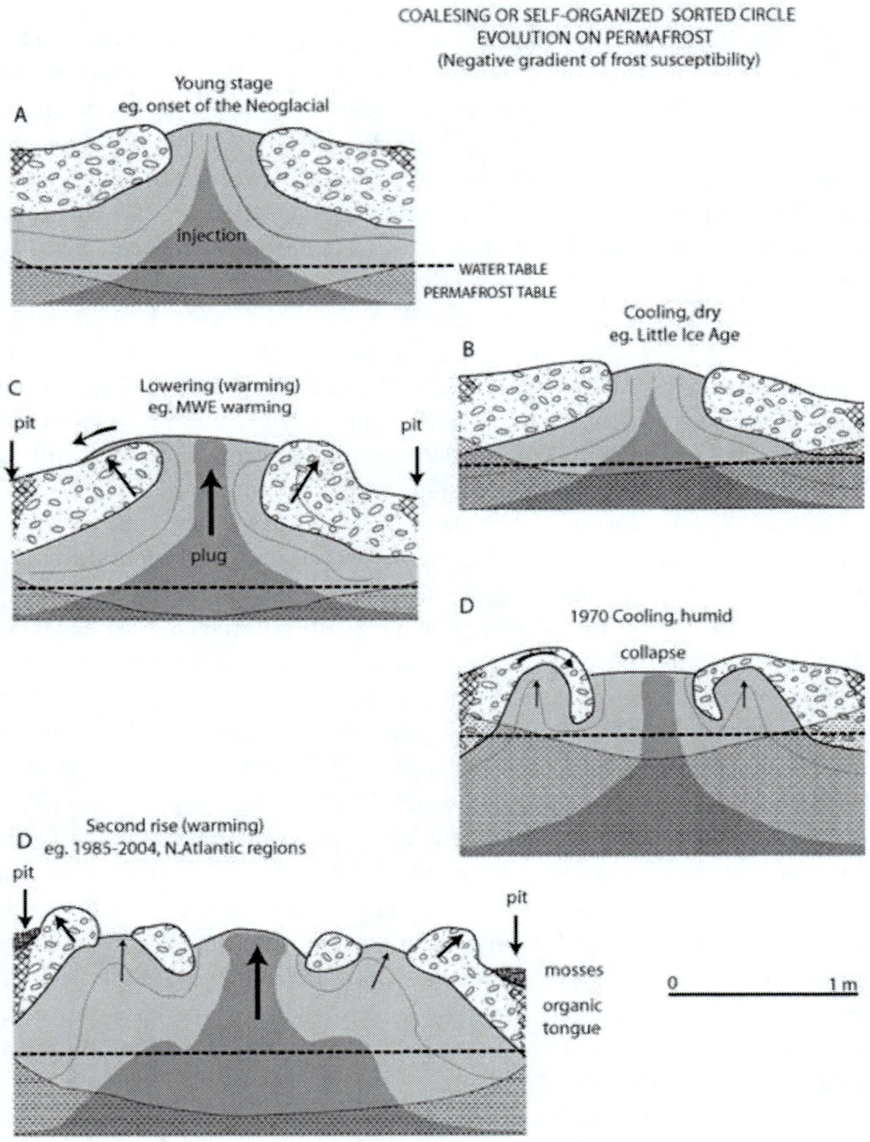

Figure 19a. Complexe evolution of sorted circles, Brögger peninsula, Svalbard.

Figure 19b. Complexe evolution of sorted circles, Brögger peninsula, Svalbard; the pictures correspond to the various steps on the figure. b) brandapynten: reactivation of the rim (1998); c) Brandalpynten,: emergence of a plug (2004) c) Stuphallet: secondary boils (light colour) on the rim (1998).

(photo G. Frost).

Figure 20. Impact of a lowering of the water-table, due to a deepening of the permafrost table resulting in successful birch seedling (NW Siberia).

This is the case for the climate of Svalbard since 1912, with a modest rise in precipitation and winter temperature (Humlum et al., 2011). This situation mostly favours vegetation extending and pedogenesis as in the Advent valley (Longyearbyen, Svalbard).

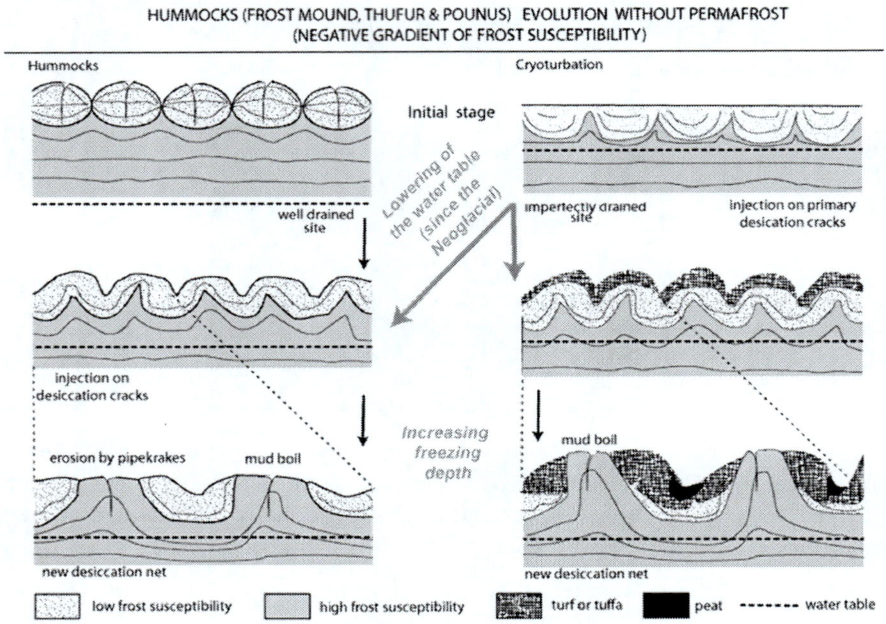

Figure 21. Impact of a lowering of the water-table and a cooling (deeper frost penetration) on earth hummocks like thufurs or pounous, negative gradient of frost susceptibility and no permafrost (modified from Van Vliet-Lanoe et al. 1998 and 2002).

If a cooling proceeds, as during the cool of 1970, the active layer becomes shallower but waterlogged in late summer. The deformations are limited by a superficial rigidity (limited differential frost heave) due to the ground saturation in ice: cryoturbation develops. Flooded patterned ground below a lake surface generally argues for dryer and colder conditions as existed during the Little Ice Age in Northern Québec. Cooling induces prostration of vegetation to escape frost injuries below the snow cover. It results in a thinner winter snow cover, a poorer thermal insulation from the ground with deeper frost penetration and lesser moisture available in springtime for the secondary frost heave below the melting ground. Cooling is also associated with increasing wind intensity, especially in winter, and a patchy snow cover because of deflation, favouring desiccation cracks and mostly thermal cracking. In loesses, a shift to hummocky microrelief marks the transition with more arid and massive accumulations; it is the case of these of the Advent valley (central Svalbard) or of the Weichselian Pleniglacial in European Armoric. Drainage changes related to cooling and to modification vegetation result in rises of the water-table within depressions (with or without permafrost) and increased frost penetration at depth, which may be reflected in changes in horizon geometry within a soil catena. Patterned grounds may thus develop on soils with former limited frost penetration.

Earth hummocks found outside the southern limit of present-day permafrost are believed to have formed during a previous cooler climate when the area was underlaid by permafrost (Tarnocai & Zoltai, 1978; Washburn 1979). In cryoturbated sub-Arctic podsols or andosols (frost mounds), an upward injection of the spodic/andic horizon (Figure 21) may result, as observed in the range of extent of permafrost: in Lapland (sporadic permafrost; Van Vliet-Lanoë et al., 1993) and Northern Québec (discontinuous permafrost, Van Vliet-Lanoë, 1997). It is also observed outside the range of continuous permafrost, in Québec (Brown & Gangloff, 1981), or Northern Iceland (Van Vliet-Lanoë et al., 1998), in the Southern Alps from 1500 m (Van Vliet-Lanoë, 1997), in the mountains of Northern Ireland (Wilson & Sellier, 1995), and the Sierra Nevada in Spain (Gomez et al., 2001; Tanarro et al., 2001), creating or reactivating patterned ground.

This signals the Neoglacial cooling (from ca 5000/4500 yrs BP, Nesje & Kvamme, 1991) and the retreat of the forest line (Arno & Hammerly, 1993; Kullman, 1995; Tinner & Theurillat, 2003) with some re-extent of permafrost, even larger that it present-day position (e.g. Thorarinsson,1964). It may be accelerated by forest fire, destroying the forest cover like in Northern Québec (Arsenault & Payette, 1992), promoting thermal cracking during the Little Ice Age (Van Vliet-Lanoë, 2005). In the permafrost regions of North America, most of the earth hummocks originated between 5000 and 2500 BP (Dyke & Zoltai, 1982), or with the Little Ice Age (Arsenault and Payette, 1993; Cook-Talbot, 1991), and in Svalbard many forms developed after 2000 BP (Van Vliet-Lanoë, 1997) and even more recently. Icelandic thufur growth started around 4500 BP and was effective during four other episodes of the late Holocene (Van Vliet-Lanoë et al., 1998). Most thermokarst features occur at the southern fringe of the discontinuous permafrost aggradation during the Little Ice Age or the 1970ties cooling (Osterkamp & Romanovsky, 1999).

Some of the patterned grounds may be inherited initially from the Last and/or Late Glacial (Ballantyne & Harris; 1994, Washburn, 1979), further transformed by Holocene pedogenesis, which accentuated frost-susceptibility gradients. South of the present-day extent of permafrost, they may be (re)activated by the Little Ice Age cooling and by anthropogenic practice modifying the vegetation cover (forest clearance and soil erosion) as in Québec, in southern Sweden (Huskvarna region), in Iceland (Orradottir et al., 2008; Schunke, 1977) or on the Cause Méjan in southern France (Joly & Dewolf, 1983).

CONCLUSION

Patterned grounds mostly result from differential frost heave, controlled by moisture supply, in other words the level of the water-table at the onset of the freezing season. They may develop as well on permafrost, as outside its range deep frost is the necessary condition. Supplementary contributing processes are contrast and gradient in frost susceptibility, desiccation cracking, thermal cracking, frost shattering, frost jacking, frost creep and pedogenesis (particle translocation). Their repartition is strongly controlled by the vegetation cover, its capacity to retain snow and its insulating properties when dry. Once initiated, patterned grounds may survive for millennia if they are not disturbed by glacial extent or a marine transgression like at the Kwadehuk Peninsula (Svalbard); they adapt to the various thermal and moisture condition changes, and are sometimes inactivated by ageing. Their story may be strongly polycyclic. These forms may develop also on slopes, stretched by frost creep, evolving into ground stripes or steps, following the available moisture during the thawing. Biological activity, especially in mountain areas, may adapt to these changes in drainage and frost heave leading to specific vegetation morphologies. In this case, patterned ground may be partly biogenic as in temperate mountains.

ACKNOWLEDGMENTS

This work is based on field experience from 1981 in Svalbard, Northern Québec (1994), Ungava (1992), and Lapland thanks to the funding of the French CNRS GDR Arctique, GDRE Arctique and also the French Polar

Institut P. E. Victor (IPEV) for Iceland and the CRYOCLIM programme in Svalbard (2004). We express our gratitude to our Colleagues Matti Sëppälä (Helsinki, Finland), Serge Payette (CEN Québec) and Michel Allard (CEN Québec) for allowing us to participate in field researches in the Arctic and sub-Arctic. We are also grateful to Albert Pissart (Liège, Belgium) and Link Washburn (Seattle) for long discussions and idea sharing. We also thank Jean-Pierre Coutard for many hours during the nineties in the laboratory of Caen (France), and building and running experiments in cold rooms. A great thanks also to Albert Pissart, Jean-Claude Dionne and Gerald V. Frost for providing their field pictures.

REFERENCES

Allard M. and Kasper J., 2001. Holocene climatic changes as detetected by growth and decay of ice wedges on the southern shore of Hudson Strait, northern Québec , Canada. *Holocene* 11, 563 578.

Arno, S.F., and Hammerly, R.P., 1993. *Timberline - Mountain and Arctic Forest Frontiers. Seattle.* The Mountaineers, 304 pp.

Arsenault D. and Payette S. 1992. A postfire shift from lichen-spruce to a lichen-tundra vegetation at tree-line. *Ecology*, 73, 3:1067-1081.

Ballantyne, C. K. & Matthews, J. A. 1982, The development of sorted circles on recently-deglaciated terrain, Jotunheimen, Norway. *Arct.Alp. Res.*. 14, 341-354.

Ballantyne, C. K. & Harris, C. 1994. *The Periglaciation of Great Britain* Cambridge Univ. Press. 330 p.

Benedict J.B., 1976 : Frostcreep and gelifluction features: a review. Quat Res., 6 : 56-67.

Beskow, G., 1935. Tjälbildningen och tja¨llyftningen med sa¨rskild ha¨nsyn till va¨gar och jarna¨gar. Sver. *Geolog.Unders. Arsbok* C 26, 375.

Black, P. B. 1995. RIGIDICE model of secondary frost heave, *CRREL Tech. Rep.* 95-12, U.S. Army Cold Reg. Res. and Eng. Lab., Hanover, N. H.

Black, P. B., and R. D. Miller 1985. *A continuum approach to modeling frost heave*. Am. Soc. of Civ. Eng., New York.

Brown, J-L and Gangloff, P., 1980. Géliformes et sols cryiques dans le sud de l'Abitibi, Québec, *Géogr. Phys. Quat.*, 34, 137-158.

Butzer K., 1971.-*Recent history of an Ethiopian delta.* Research Paper, Department of Geography , University of Chicago, 136, 184p.

Cegla J. and Dzulinski S., 1970 Uklady niestatcznie wartwane I ich wytepowanie w srodowisku peryglacjalny. (System with reversed density gradient and their occurrence in peryglacial zones). *Acta Univ.Wratislaviensis*, 124, 17-42.

Cook-Talbot, J. 2013. Sorted circles, relative-age dating and palaeoenvironmental recons-truction in an alpine periglacial environment, eastern Jotunheim, Norway: lichenometric and weathering-based approaches. *Holocene,* 1, 128-141.

Corte, A., 1966. Particle sorting by repeated freezing and thawing. *Biul. Peryglac.*, 15, 176-240.

Coutard, J.P., Van Vliet-Lanoë, B., Auzet, A.V., 1989. Experiments concerning the genesis and the deformation of soil structures by frost creep. *Z. Geomorph.* NS 71, 13–23, Suppl.

Crampton, C. B. ,1977,. A study of dynamics of hummocky microrelief in the Canadian north. Canadian J.Earth Sci., 14, 639-649.

Dionne, J.C., 1974. Cryosols avec triage sur ravage et fond de lacs, Québec central subarctique. *Rev.Geogr.Montréal*, 28,323-342.

Dyke, A. S. and Zoltai, S. C. 1982. Radiocarbon dated mud-boils, Central Canadian Arctic. In Proc.VI Inter.Conf. Permafrost, Fairbanks. NRC, pp. 271-274.

Dylikowa, A.1961 Structures de pression congélisatiques et dstructur de gonflement par le gel de Karatzynow près de Lódz. *Bull.Soc.Sci. Let., Lódz*, 12, 9, 1-23.

Dżułnsky, S. and Walton, E.K., 1963. Experimental production of sole markings. *Geol.Soc.Edinburgh Trans.*, 19, 279-304.

Egginton, P.A., 1981. The impact of disturbance on mudboil activity, north Henik Lake, District of Keewatin. In Current research, part A. Geological Surv. Canada, paper 81-1A, 299-303.

Eisseman, L., 1981. Periglaziäre Prozesse und Permafroststrukturen aus sechs Kaltzeiten des Quartars. *Altenberger Naturwiss. Forsch.* 1, 171 pp.

Etzelmüller B., Sollid J.L. 1991. The role of weathering and pedological processes for the development of sorted circles on Kvadehuksletta, Svalbard - a short report *Polar Res.* 9, 181–191.

Forman, S. and Miller, G.H., 1984. Time-dependent soil morphologies and pedogenetic processes on raised beaches, Bröggerhalvöya, Spitsbergen, Svalbard. *Arct. Alp. Res.,* 16, 381-394.

Fowler, A. C. 1989. Secondary frost heaving in soils, SIAM *J. Appl. Math.*, 49, 991– 1008, doi:10.1137/0149060.

French H.M., 1986: Periglacial involutions and mass displacement structures, Banks Island, Canada. *Geografiska An.,* 68A, 3: 167-174.

Frenot, Y., Van Vliet-Lanoë B. and Gloaguen J.C , 1995 - Particle translocation, cryogenic processes and soil development on a glacier foreland, Iles Kerguelen. *Arct.Alp. Res.* ,27, 107-115.

Frost G.V., Epstein H.E, Walker D.A., Matyshak G. & Ermokhina K. 2013, Patterned-ground facilitates shrub expansion in Low Arctic tundra *Environ. Res. Lett.* 8 015035 doi:10.1088/1748-9326/8/1/015035.

Gilpin, R. R., 1980. A Model for the Prediction of Ice Lensing and Frost Heave in Soils, *Water Resour. Res.,* 16, 918–930.

Glen, L.W., 1974. The physics of ice. US Army CRREL Rep., II-C2a, 81pp.

Gomez, A., Palacios, D., Ramos, M., Tanarro, L.M., Schulte, L., Salvador, F., 2001. Location of permafrost in marginal regions: Corral del Veleta, Sierra Nevada, Spain. *Permafrost Periglac. Proc.* 12, 93–110.

Hallet, B., and Prestrud, S., 1986., Dynamics of periglacial sorted circles in western Spitsbergen, *Quat. Res.,* 26, 81–99.

Hallet, B., and Waddington, E. C., 1992. Buoyancy forces induced by freeze-thaw in the active layer: implications for diapirism and soil circulation, in *Periglacial Geomorphology*, edited by J. C. Dixon and A. D. Abrahams, , John Wiley, Chichester, U.K pp. 251– 279.

Harris, C., M. C. R. Davies, and B. R. Rea (2003), Gelifluction: Viscous flow or plastic creep?, *Earth Surf.Proc.Land.,* 28,1289-1301.

Herz ,K., and Andreas G., 1966. Untersuchungen zur Morphologie de periglazialen Auftauschicht im Kongsfjordgebiet (Westspitzbergen) *Petermans Geogr. Mitteilungen*, 110, 190-198.

Hoekstra, P.,1966, Moisture movement in soils under temperature gradients with the cold-side temperature below freezing. *Water Resources Res.*, 241-250.

Humlum, O., Solheim, J.-E., and Stordahl, K. 2011. Identifying natural contributions to late Holocene climate change. Global Planet. Change 79, 145-158.

Jacobsen ,J., 1989. Evidence of translocation into the B horizon of a subarctic Podzol in Greenland. *Geoderma*, 45: 3-17

Joly F. and Dewolf Y.1983. Le causse Méjean : introduction géologique et géomorphologique. Paris VII, Université, 8p...

Kade, A., and Walker, D. A., 2008. Experimental alteration of vegetation on nonsorted circles: Effects on cryogenic activity and implications for climate change in the Arctic. *Arct. Antarct. Alp. Res.*, 40, 96–103.

Kaplar, C.W., 1974. Phenomena and mechanism of frost heaving. Highway Res.Rec., 304:1-13.Van V Locke, W. ,1986. Fine particles translocation in soils developed on glacial deposits, Southern Baffin Island, NWT, *Canada. Arct. Alp. Res.,* 18, 33-43.

Kessler M. A., Murray A. B., Werner B. T., Hallet B., 2001 A model for sorted circles as self-organized patterns. *J. Geophys. Res.*, 106, 13287-13306.

King, L., 1986. Zonation and ecology of high mountain permafrost in Scandinavia. *Geografiska An.* 68A (3), 131–139.

Kuenen, P.H. 1958. Experiments in geology. *Trans.Geol.Soc. Glasgow*, 23: 1-28.

Kühle, M.,1983. Permafrost and Periglacial Indicators on the Tibetan Plateau from the Himalayan Mountains on the South to the Quilian Shans on the North. In: 4th Int. Conf., Permafrost, Fairbanks, Alaska, 1983 Univ. Fairbanks, 159-160.

Kullman, L., 1995. Holocene Tree-Limit and Climate History from theScandes-Mountains, Sweden. *Ecology,* 76, 2490-2502.

Lachenbruch A.H., 1966. Contraction theory of ice wedge polygons: a qualitative discussion. Permafrost Intern. Conf.Proc.Nat.Acad.Sc. and Nat.Res.Council USA publ. 1287, 63 71.

Lautridou, J.P., 1985: Le cycle périglaciaire pléistocène en Europe du Nord-Ouest et plus Public. Centre de Géomorphologie CNRS, Caen, 907 pp.

Lewkowicz A.G., 1994.. Ice wedge rejuvenation, Fosheim peninsula, Ellesmere Island. *Permafrost Periglac. Proc.*, 5 : 251 268.

Lindner,L., 1976. Frostweathering of the pre quaternary rocks and its influence on landscape evolution in the northwestern margin of the HolyCross Moutains. *Acta Georgr. Lodziensia*, 71,149-164.

Locke, W. ,1986. Fine particles translocation in soils developed on glacial deposits, Southern Baffin Island, NWT, Canada. *Arct. Alp. Res.*, 18, 1 : 33-43.

Lull, H.W. and Rushmore, F.M. 1961. Influence of forest cover on snow and frost in the Adirondaks. Eastern Snow.Conf.Proc. 18th Ann.Meet., 69-71.

Mackay, J.R., 1980. The origin of hummocks, western Arctic coast, Canada. *Can.J.Earth Sc.,*17: 996-1006.

MacRoberts, E.C. 1978. Slope Stability in cold regions. In Andersland O.B. and Anderson D.M. (eds): *Geotechnical engeneering for cold regions*, Mac Graw Hill, New York, pp.363- 404.

Mériaux, S., 1979. Le sol et l'eau. In: Bonneau, M., Souchier, B. s.., Pédologie II, Constituants et propriétés du sol. Masson, Paris, pp. 266–312.

Miller, R.D.,1972. Freezing and heaving of saturated and unsaturated soils. In "Highway Res. Board, Frost action in soils." Nat. Acad. Sci. and Eng. Highway Res. Rec., 393: 1-11.

Murton, J.B., Peterson, R. & Ozouf, J.C 2006. Bedrock Fracture by Ice Segregation in Cold Regions". Science 314, 1127–1129.

Murton, J.B and French, H. M. ,1993. Thermokarst involutions, Summer Island, Pleistocene Mackenzie Delta, western Canadian Artic. *Permafrost Periglac. Proc.*, 4. 217-229.

Nesje A. & Kvamme,M, 1991 Holocene glacier and climate variations in western Norway: Evidence for early Holocene glacier demise and multiple Neoglacial events. *Geology*, 19, 610-612.

Orradóttir B., Archer S.R., Arnalds; Ó., L.P. Wilding and T.L. Thurow. 2008. Infiltration in Icelandic Andisols: The role of vegetation and soil frost. *Arct.Antarct.Alp.Res*. 40:412-421.

Osterkamp, T.E.and Romanovsky, V.E., 1999. Evidence for warming and thawing of discontinuous permafrost in Alaska. *Permafrost Periglac. Proc.* 10, 17–37.

Paton, T.R., 1974. Origin and terminology for gilgaï in Australia. Geoderma, 11, 221-42.

Payette, S., Gauthier, L. and Grenier ,Y., 1986. Dating ice-wedge growth in subarctic peatlands following deforestation. *Nature,* 322, 6081: 724-727.

Penner E. 1986. Ice lensing in layered soils. *Can. Geotech. J.* 23, 334 – 340.

Penner E., and Goodrich L.. ,E. 1981. Location of segregated ice in frost susceptible soil. *Engineering Geology* (Amsterdam), 18, 231-244.

Peterson, R. A., and. Krantz, W. B, 2003. A mechanism for differential frost heave and its implications for patterned ground formation, *J. Glaciol.,* 49,69– 80.

Peterson, R. A.; Krantz, W. B., 2008. Differential frost heave model for patterned ground formation: Corroboration with observations along a North American arctic transect. *J. Geophys. Res.*113: G03S04.

Pissart, A., 1982. Déformation de cylindres de limons entourés de graviers sous l'action d'alternances el-dégel. *Biul. Periglac*., 29, 219-229.

Pissart A., 1976. Sols à buttes, cercles non triés et sols striés non triés de l'Ile de Banks, (Canada, NWT). *Biul. Periglac*.26,275-285.

Pissart, A.,1969. Le mécanisme périglaciaire dressant les pierres dans le sol. Résultats d'expériences. *C.R. Acad. Sc. Paris*, 268 : 3015-3017.

Pissart, A., 1966. Expériences et observations à propos de la genèse des sols polygonaux triés. *Rev.belge Géogr.,* 90,1-19.

Rempel, A.W., 2007. "Formation of ice lenses and frost heave". *J.Geophys.Res.*112 (F02S21).

Schunke, E. 1977. Zur genese der Thufur Islands und Öst Grönlands. *Erdkunde,* 31, 279-287.

Schunke E. and Zoltai, S.C., 1988. Earth Hummocks (thufur). In M. J. Clark edit.: *Advances in Periglacial Geomorphology.* J. Wiley , 231-245.

Sharp, R.P.1942. Periglacial involutions in northeastern Illinois. J. Geology, 50,113-133.

Shilts, W.W.,1978. Nature and genesis of mudboils, central Keewatin, Canada. Canadian *J.Earth Sci.,* 10:1053-1068.

Sigafoos, R. S., 1951. Soil instability in tundra vegetation. *Ohio J. Sci.,* 6: 281–298.

Solovyev, P.A., 1973. Thermokarst phenomena and landforms due to frost heaving in central Yakoutia. *Biul.Peryglac.* 23, 135–155.

Svensson, H., 1973. Distribution and chronology of relict polygon patterns on the Laholm plain, the Swedish west coast, *Geografiska An.,* 54-A (3-4), 159-175.

Svensson, H. 1977. Observations on polygonal fissuring in non permafrost areas of the Norden Countries. IN H.Poser edit. Formen, Formengesellschaften und Untergrenzen in den heutigen periglazialen Hëhenstufen der Hochgebirge Europas und Afrikas zwische Arktis und Equator. Abh. Akad. *Wiss. Göttigen, Math Phys.* Kl., 3(31), 63-67.

Taber, S., 1929. Frost heaving. *J..Geology,* 37, 428-461.

Tanarro, L. M., Hoelzle, M., García, A., Ramos, M., Gruber, S., Gómez, A. Piquer, M. and Palacios, D. 2001. Permafrost Distribution Modeling in the Mountains of the Mediterranean: Corral del Veleta, Sierra Nevada, Spain. *Norwegian J. Geogr.* 55, 253-260.

Tarnocai, C. and Zoltai, S., 1978. Earth hummock of the canadian Arctic and Subarctic. *Arct. Alp. Res.*, 10, 581-594.

Tedrow, J.C.F., 1977. *Soils of the polar landscape.* New Brunswick, N.J.: Rutgers Univ.Press., 638p.

Tinner, W. and Theurillat, J.-P., 2003. Uppermost limit, extent, and fluctuations of the timberline and treeline ecocline in the Swiss Central Alps during the past 11,500 years. *Arct., Antarct., Alp. Re*s., 35,158-169.

Thorarinsson, S. 1964. Additional notes on patterned ground in Iceland with particular reference to ice-wedge polygons, *Biul. Peryglac.* 14 327-336.

Tystovich, N. A. 1975. *The Mechanics of Frozen Ground*, McGraw-Hill, New York. 426 pp.,

Van Vliet-Lanoë, B. 1998. Pattern ground, hummocks and Holocene climate changes. *Eurasian J. Soil Sc.,* 31, 507-513.

Van Vliet-Lanoë B. 1992: Solifluction activity in N-Atlantic Arctic and Subarctic: the microstratigraphical record of changing climate during the second half of the Holocene. *Palaeoklimatfors.,* 6, spécial ESF Proj. 279-294.

Van Vliet-Lanoë, B., 1991. Differential heave, load casting and convection: converging mechanisms. A discussion of the origin of cryoturbations. *Permafrost Periglac. Proc.*, 2 , 123-139.

Van Vliet-Lanoë B. 1988a : The genesis of cryoturbations and their significance in environmental reconstruction. *J. Quat. Sc.* 3, 85-96.

Van Vliet-Lanoë, B., 1988b: The origin of patterned grounds in N.W. Svalbard, V. Int. Conf. Permafrost, Trondheim, Norway. Tapir Publishers 2: 1008-1013.

Van Vliet-Lanoë B. 1988c: *Le rôle de la glace de ségrégation dans les formations super-ficielles de l'Europe de l'Ouest. Processus et héritages.* Thèse de Doctorat d'Etat, Géomorphologie, Université de Paris I, 1987, Editeur Editec, Caen, 1988 - 854p.

Van Vliet-Lanoë, B., 1985. Frost effects in soils. In J. Boardman (editor)"S*oil and Quaternary Landscape Evolution*", J. Wiley Publ., pp 115-156.

Van Vliet-Lanoë B. 1983 : *Etude cryopédologique au S. du Kongsfjord, Svalbard. Rapport de la mission Spitzberg*, 1982. Public. interne Centre de Géomorphologie CNRS, offset, 40p.

Van Vliet-Lanoë B., Bourgeois O., and Dauteuil O. 1998. Thufur formation in Northern Iceland and its relation to Holocene climate change. *Permafrost Periglac. Pro*c., 9. 347-365.

Van Vliet-Lanoë B., Coutard J.P. and Pissart A, 1984: Structures caused by repeated freezing and thawing in various loamy sediments. A comparison of active, fossil and experimental data. *Earth Surface Proc. Landf.,* 9: 553-565.

Van Vliet-Lanoë B., Maygari A., Meilliez F., 2004: Distinguishing between tectonic and periglacial deformations of quaternary continental deposits in Europe. *Global Planet. Change* 43, 103-127.

Van Vliet-Lanoë B. & Seppälä M., 2002: Stratigraphy and Formation of Pounus (Frost Mounds) in Finnish Lapland. *Holocene,* 12, 187-199.

Van Vliet-Lanoë, B., Seppälä, M. and Käyhkö, J.,1993. Dune dynamics and cryoturbation features controlled by Holocene water level change, Hietatievat, Finnish Lapland. *Geologie en Mijnbouw ,* 2, 211-224.

Vandenberghe J., 1992. Cryoturbations: a sediment structural analysis. *Permafrost Periglac.Proc.* 3, 343 352.

Walker, D.A., Billings, W.D., de Molenaar, J.G. 2001. Snow-vegetation interactions in tundra environments. in H.G. Jones, R.W. Hoham, J.W. Pomeroy, D.A. Walker (eds.) *Snow Ecology*. Cambridge Univ Press.. pp. 266-324.

Wang, B. and French Hugh M. 1995 Implications of Frost Heave for Patterned Ground, Tibet Plateau, China. *Arctic Alpine Res.* 27,. 337-344.

Wang, B. & French, H.M., 1994 Climate controls and high-altitude permafrost, Qinghai-Xizang (tibet) Plateau, China *Permafrost Periglac.Proc.* 5 87–100.

Washburn, A.L., 1997. Plugs and plugs circles. A basic form of patterned ground, Cornwallis Island, Arctic Canada. Origin and implications. *Geological Soc. Amer.memoir* 190, 87 p.

Washburn A.L., 1980: Permafrost features as evidence of climate change. *Earth Sc.Rev.*, 15:327-402

Washburn, A.L. 1979. Geocryology. A survey of periglacial processes and environment. Arnold Publ. London, 406 pp.

Washburn, A. L., 1956. Classification of patterned ground and review of suggested origins, *Geol. Soc. Am. Bull.*, 67, 823– 865.

Washburn, A.L., Mima Mounds, 1988. An Evaluation of Proposed Origins with Special Reference to the Puget Lowlands. Washington: Washington Division of Geology and Earth Resources Report of Investigations 29,

Washburn, A.L., Smith D.D., Goddard R.H., 1963. Frost cracking in a middle latitude climate. *Biul. Periglacial.*12, 175-189.

Williams, P.J. and Wood, J.A. 1982. Investigations of moisture movements and stresses in frozen soils. Final report of the Geotechnical Sciences Laboratoires, Carleton University, Ottawa, Contract serie n°OSV81-00119.

Wilson P. and Sellier D., 1995 - Active periglacial patterned ground and cryoturbation on Muckish Mountain, Co. Donegal, Ireland, *Permafrost Periglac.Proc,* 6, pp. 15-25.

Woo M.K., Heron R. & Marsh P.,1982: Basal ice in high Arctic Snowpacks. *Arct Alpine Res* 14 (3), 251-260.

Zoltai, S.C.,1973: Vegetation, surficial deposits and permafrost relationships in the Hudson Bay lowland. Proceedings Symposium Physical Environment of the Hudson Bay Lowland, Univ. Guelph (Ont.), p. 17-39, 9 Figure.

In: Permafrost
Editor: Oleg S. Pokrovsky
ISBN: 978-1-62948-830-1

Chapter 3

ZONE OF OPTIMUM DEVELOPMENT OF PATTERNED GROUNDS AS A CHARACTERISTIC OF SEMI-ARID AREAS, AN INDICATOR OF PERMAFROST-LINES AND A REFERENCE TO SUB-GLACIAL PERMAFROST BELOW THE PAST ARID COLD-BASED TIBETAN ICE SHEET

Matthias Kuhle*
Geography and High Mountain Geomorphology,
Geographisches Institut der Universität Göttingen,
Goldschmidtstr. Göttingen/Germany

ABSTRACT

Since 1970 the results summarized here have been gathered by field researches carried out in the Arctic (Spitsbergen (Svalbard), Greenland and Alaska) as well as in Tibet and in the high mountains on the North- and South-hemisphere. Now the publication of the field data takes place with a synthetic evaluation of all findings.

* Address for correspondence: Prof. Dr. M. Kuhle – *E-mail: mkuhle@gwdg.de.*

1. The global uniformity of permafrost- and periglacial indicators is evidenced by the fact that they do not only occur in high arctic latitudes, but also in subtropical mountains. Pingos as well as polygonal frost cracks have been met between 28 and 35°N above of c. 4600 - 5300 m asl. They are considered to be key forms of arctic periglacial as well as permafrost areas. Thus it can be proved that the planetary (zonal) periglacial change of forms is repeated hypsometrically in the direction of the equator.
2. In a reverse symmetrical way as the arctic periglacial store of forms with key forms like patterned grounds is completely situated in the permafrost area, in the direction of the equator the periglacial store of forms with its patterned grounds can be found everywhere in the constantly thawed soil, namely where the mountain summits just do not reach the permafrost limit anymore. In a systematic difference to the Arctic, where the lower border of permafrost not even is reached from above, numerous mountains of the temperate, the subtropical and the tropic latitudes not only reach the lower limit of the permafrost from below, but tower above it by up to several 1000 m. Whilst the concerning tropical mountains due to their humidity are glaciated above the permafrost limit, the subtropical mountains are showing patterned grounds up to far above the permafrost line (c. 500 – 900 altitude meters). Due to their portion of precipitation that is situated in between, the moderate humid mountains necessarily hold an intermediate position. Being free of glaciers they thus tower above the permafrost line by only a few 100 meters.
3. In the semi-arid to arid subtropical high mountains the size of the individual forms of patterned grounds (i.e., their diameter and the depth of their sorting of material) does not only decrease in the area of the ever-thawing-soil with diminishing absolute altitude towards below- as this applies for all extra-arctic mountains – but also decreases in the area of the permafrost soil with increasing absolute altitude towards above.
4. This finding proves sort of vertical mirror symmetry of the periglacial altitude level in the approximate centre of which the optimum zone of development of patterned ground is situated.
5. Our investigations showed that important parts of the nearly 2 million km² extended ground moraines of the glacial Tibetan ice sheet have been formed under subglacial permafrost conditions. This speaks in favor of the past existence of a cold-based inland ice. Mainly, however, the persistence of permafrost is proved under the more than 1000 m-thick sheet of ice. Here a glacial phenomenon is concerned, that currently in all probability nowhere exists. The most probable is its actual realization in the marginal areas of the Antarctic.

6. Currently there can be differentiated 7 types of combinations of periglacial levels with patterned grounds and glacial levels on the globe.

1. Global Uniformity of Permafrost- and Periglacial Indicators

To provide evidence that there exists a global uniformity of permafrost- and periglacial indicators, these indicators must be met in high latitudes, i.e., in the arctic, as well as in low latitudes, i.e., in the mountains as far as to the equator. Patterned grounds have been proved for the arctic zone (Cailleux & Taylor 1954; Svensson 1967; Tricart (ed.) 1967; Washburn 1956, 1979) as far as to the subtropical and tropical high mountains (Ball & Goodier 1970; Brosche 1971; Fränzle 1959; Graf 1973; Hastenrath 1973; Hövermann 1972; Hövermann & Kuhle 1978; Klaer 1977; Kuhle 1978a, b, 1983a, b; Rathjens 1964; Troll 1947). First there was the opinion that there could be found clearly larger patterned grounds in the arctic (Troll 1944), in the sorting of which also larger boulders were integrated.

Photo 1. At c. 500 – 400 m asl on the old face in the N-slope of the Nordenskjöldfjellet (W-Svalbard: 78°13'7.19"N 15°27'1.57"E, c. 460 m asl) patterned grounds developed in Late-Glacial to Holocene ground moraine. The individual forms are 1-2 m in diameter; for comparison of the size: the length of the ice axe is 55 cm. Analogue photo M. Kuhle, 29/07/1976.

Photo 2. Patterned grounds of the arctic type; the singular forms are 2-3 m in diameter, the crystalline boulders integrated into the sorting are meter-sized; locality: N-side of Galdhöpiggen (61°40'27.52"N 8°21'56.56"E; c. 1850 m asl) at the Juvvatnet in the Scandes, Norway, N-Europe. Analogue photo M. Kuhle, July 1971.

However, due to the increasing knowledge about subtropical mountains this has been disproved. So for instance in Spitsbergen (Photo 1), in the southern Skandes (Photo 2) as well as in the Andes at 32°S about 4440 m asl (Photo 3) and in the Tibetan Himalaya at 28-30°N about 4800 m asl there occur just as large patterned grounds of up to 2-3 m in diameter (Kuhle 1982a: Abb. 72) (Figure 1). This applies, too, for Tibet where similar patterned grounds (Photo 4) and- on sloping faces of ground moraine - also large stripes of boulders, that means also forms in arctic dimensions, have been observed (Photo 5) (Kuhle & Roesrath 1990: 99). For this reason those key forms that for a long time have been restricted as belonging to the Arctis, with regard to the global unity of the periglacial inventory of forms are still more convincing. These are pingos (Photo 6) (Svensson 1970) and frost crack nets with tundra polygons (Tundren Polygone), i.e., nets of ice wedges ("Eiskeilnetze") (cf. Rapp & Annersten 1969; Washburn 1979) (Photo 7). However, in fact permafrost indicators like these as well as permafrost itself (Photo 8) can also be met at great heights of subtropical latitudes (Photos 9, 10; Figure 7) (Kuhle 1983c, 1985a). Further permafrost indicators are rock glaciers. They also occur in the Arctis, as e.g. in Alaska (Photo 11) (cf. Wahrhaftig & Cox 1959), as well as in subtropical mountains (Grötzbach 1965; cf. also Boesch 1951) like the Andes, e.g. in the Aconcagua-massif at 32°S (Photo 12).

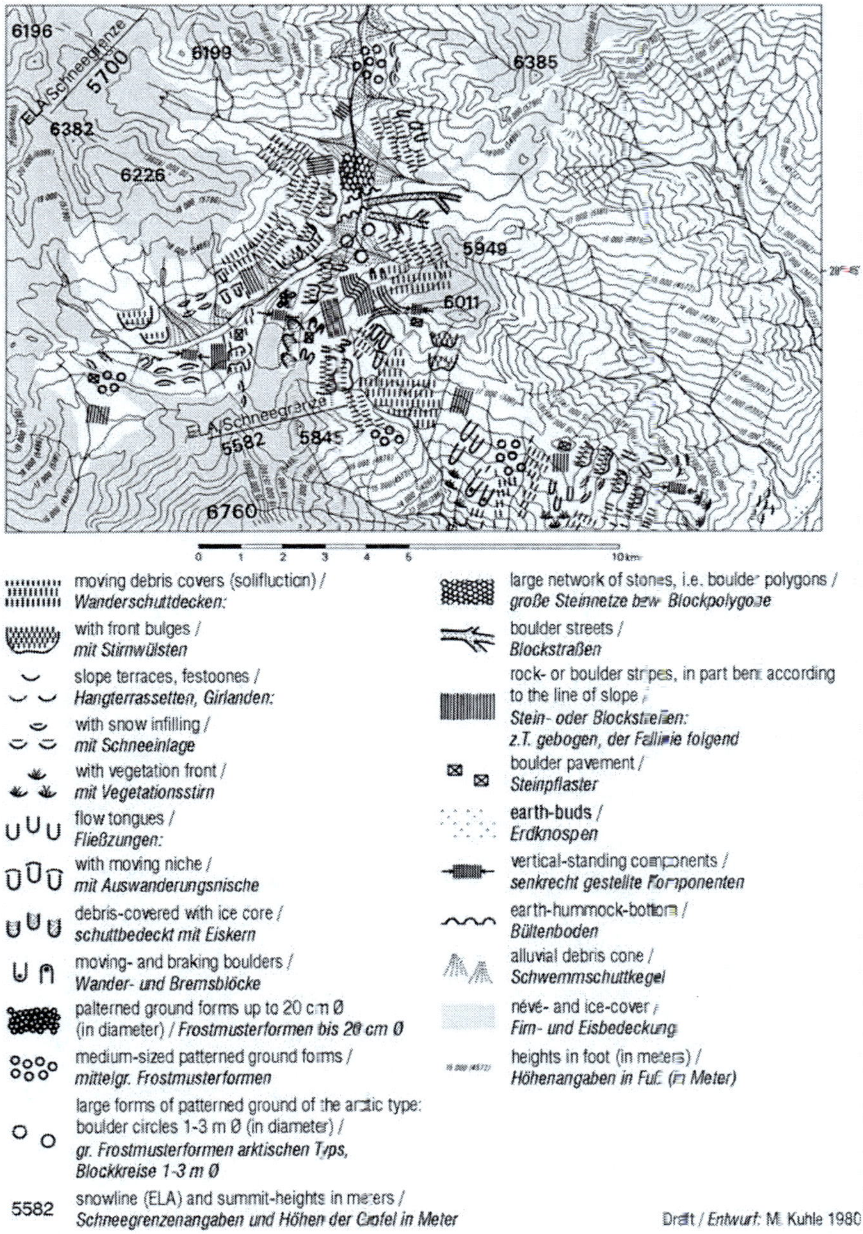

Figure 1. Periglacial forms in the ‚Hidden Valley'; the forms of patterned ground and their arrangement are representative of the high valleys of the Inner (Tibetan) Himalaya, north of the Dhaulagiri-group (28°20'-28°55'N/83°-84°E).

Photo 3. Patterned grounds with large individual forms of 1-2 m in diameter – here exactly 1.7 m - in ground moraine of trachyt debris (32°39'18.39"S 70° 4'15.27"W, 4400 m asl aneroid measurement) on the W-side of Cerro Aconcagua and SE-side of Cerro Cathedral (Co. Bonete) in the subtropical semi-arid Andes. Length of the ski-sticks 120 cm. Analogue photo M. Kuhle, 18/02/1980.

Photo 4. Patterned grounds in arctic dimensions with 2-3 m large individual forms in the sorting of which meter-sized granite boulders have been integrated; NE-Tibet in the NE-slope of the Kakitu-massif (38°8'6.29"N 96°29'49.59"E, c. 4850 m asl). Analogue photo M. Kuhle, 06/08/1981.

Photo 5. Patterned grounds with block- i.e., earth-stripes in arctic dimensions in NE-Tibet on the plateau area of the E-Kakitu-massif (38°7'4.95"N 96°29'53.75"E, c. 4920 m asl); the individual forms of the periglacial sorting of material on the about 2-3° inclined ground moraine surface are showing extensions of several meters; the meter-sized moraine boulders, integrated into the sorting, consist of granite with crusts of ferromanganese. Analogue photo M. Kuhle, 05/08/1981.

Photo 6. Pingo: Key form of arctic permafrost in W-Greenland on the peninsula Nuqssuaq (70°3'29.61"N 52°2'53.29"W); in the SSW (background) the Fjord of Saqaq with its icebergs can be seen. The pingo is situated on a glaciofluvial delta-accretion (foreset beds) over ground moraine. It is c. 15 m high with a basal circumference of c.200 m. The surface that has subsided in a slight crater-form, points to the thawing of the ice core. Analogue photo M. Kuhle, July 1979.

Photo 7. Permafrost ice wedge (tundra-polygon) at Dicksonland in W-Spitsbergen (78°33'12.40"N 15°27'45.85"E, 250 m asl); at the time when the picture was taken, melt water stood on the ice wedges in the small trenches (⇩) of the polygonal frost cracks that fringe the tundra-polygons and differentiate them from each other. Analogue photo M. Kuhle, 29/6/1976.

Photo 8. The dug exposure of ground moraine in E-Central Tibet (34°09'40"N97°43'50"E, c. 4515 m asl, aneroid-measurement) exposes the permafrost (□) under the 1-1.5 m thick active layer (the annual thawing bottom); (•) marks far-travelled erratic granite boulders from the Bayan Har Shan in the S. Analogue photo M. Kuhle, 27/7/1994.

Photo 9. Pingo in N Central Tibet (35°40'14.06"N 94°2'59.63"E) at 4650 m asl (aneroid-measurement) in material of ground moraine (person on the right margin for comparison). (□) mark the artificially exposed injection- or segregation ice of the permafrost milieu; (▪) shows the ground moraine cover in the further environment; it contains erratic granite boulders and lies on sedimentary rocks. Analogue photo M. Kuhle, 22/7/1981.

Photo 10. Polygonal frost cracks as indicators of permafrost in the Tibetan Himalaya at c. 5420 m asl, somewhat S of the Thorong La (28°47'33.47"N 83°56'17.33"E). Analogue photo M. Kuhle, 9/9/1977.

Photo 11. Rock glacier as indicator of permafrost about 1450 m asl in the Alaska Range (Mt. McKinley-massif) E of the glacier tongue end of the Muldrow glacier facing N, taken from c. 1730 m asl down-valley (63°22'59.00"N 150°20'35.13"W). Analogue photo M. Kuhle, 13/7/1983.

Photo 12. Rock glacier as permafrost indicator that in the SSE-flank of the 4900 m-high E-satellite (on the right) of the c. 5400 m high Cerro Dollossa (Tolosa) (in the background on the left) flows down to c. 3850 m asl (32°46'19.56"S 69°59'56.09"W). Aconcagua-group in the subtropical Andes. Analogue photo M. Kuhle, 18/3/1980.

Accordingly, there has to be summarized the following: Because the permafrost is controlled by the lowest temperatures of the periglacial region, that means by an average annual temperature colder than -3 to -6°C, its existence proves the global uniformity of the periglacial region even in the low latitudes.

2. Proof of the Zonal (Planetary) and Regional (Hypsometric) Symmetry of Permafrost and Perennial Thawing Soil by Means of Patterned Grounds

Patterned grounds are showing the climate-genetic largest spread in the periglacial milieu. Because they are only dependent on the freezing and thawing of the soil and not necessarily on a permafrost table (Photo 13), they also occur – but not only – in the areas of permafrost. There they develop in the annual thawing soil above the permafrost table, i.e., in the active layer (Photo 1 and 14). Their development, however, is not dependent on the permafrost table. This, however, exactly applies - in contrast to patterned grounds - for pingos and tundra polygons.

Soils of freezing and thawing also occur outside of the permafrost region, i.e., below of the altitudinal level of permafrost (Heim 1936; Kuhle 1978b, 1985a). Accordingly, there are – as far as they are not suppressed by vegetation (cf. Huang Rongfu 1982: 51 Tab. 1) – also patterned grounds and related forms of solifluction (Photo 13). Here, their sorting of material caused by freezing and thawing primarily is due to a penetrating of frost into the upper layer of the permanent thawing soil. Diametrically there is released in the permafrost soil the sorting of material caused by a primary process of thawing in the upper layer of the permafrost soil (cf. Büdel 1959; Dukiewicz 1968; Fujii & Higushi; 1976; Ohlson 1964; Semmel 1969; Svensson 1967; Kuhle 1978a, 1983a, 1985a).

In general one may say that the annual depth of thawing of the permafrost soil, i.e., the thickness of the active layers decreases towards the high latitudes. Therefore - under comparable further conditions- decreases the size, i.e., decrease the diameters and integrated boulder sizes of patterned grounds towards higher latitudes. Most immediately this can be proved in the arctic mountains as e.g. Spitsbergen (Svalbard).

Photo 13. Patterned ground in debris of marl at c. 2450 m asl in the subtropical-arid Kuh-i-Jupar (massif) in the Zagros mountains in the SE-Iran (c. 29°59'30.48"N 57°9'56.47"E). Here, the desiccated and therefore sunk summer form is concerned. The rod is 1 m long. The sorting depth of the material of the fine ground beds amounts to c. 17 cm. Here similarly large patterned grounds are concerned as on the Nordenskjöldtoppen in Spitsbergen (cf. Photo 14). Analogue photo M. Kuhle, 30/3/1974.

Photo 14. Patterned grounds in siltstone i.e., marl debris at 1040 m asl on the summit plateau of Nordensköldtoppen (W-Spitsbergen, Svalbard: 78°10'43.28"N 15°24'10.82"E) The length of the ice axt amounts to 55 cm. The depth of the sorting of material of the fine earth beds is 15-18 cm. The photo shows the only superficially thawed, still high-frozen spring-form (cf. Photo 13).During days with continuous sun-radiation a thawing-depth of c. 16 cm is reached; this corresponds with the depth of sorting of the material. Analogue photo M. Kuhle, 29/07/1976.

The surveying of patterned grounds at the Nordenskiöldfjellet showed that the size of the centers of fine ground ("Feinerdebeete") decreases with the diminishing depth of thawing from c. 500 m up to c. 1040 m asl (compare Photo 1 and 14) (Kuhle 1978).

Here the copula is in a regular proportion of the depth of sorting of material to the diameter of the separate patterned ground of c. 1.5 : 3 (depth of sorting of material 1.5 units; diameter 3 units). There above at 1040 m asl, where the depth of thawing reaches c. 20 cm, diameters of patterned ground of 30 to 40 cm have been measured. This is confirmed by largest patterned grounds observed at sea level at Spitsbergen (close to Ny-Ålesund, Kingsbei; cf. Louis & Fischer, 1979: Bild 9).

The logical cross-check needs a decrease in size of the patterned grounds with a diminishing depth of frost penetration in the permanent thawing soil towards the lower latitudes. Thus, logically there has to be carried out the empirical counter check in the non-arctic mountains. Its condition is, that the height of the mountains renders seasonal frosts and, accordingly, freezing and thawing possible.

In the European Alps (Furrer 1965, Höllermann 1967), the Iberian mountains (Fränzle 1959; Höllermann 1967, Brosche 1971), the Apenin (Kelletat 1969), in Skandinavia (Ohlson 1964; Semmel 1969; Svensson 1967) as well as in many other mountains, like the Karakoram (Furrer 1986) etc a lower limit of patterned ground ("Strukturbodenuntergrenze") has been described and mapped (Graf 1973) (Figure 2). It is situated without exception in the ever thawing soil. At the same time it belongs to the characteristics of patterned grounds that they become smaller and smaller towards their lower limit and eventually come to an end completely. This shows at the same time their increasing size towards above.

Due to the thermal equivalence of hypsometric to planetary (zonal) decrease in temperature, this increase in size of patterned grounds in the ever thawing soil of mountains at the same time means also an increase of size in the ever thawing soil towards the higher latitudes. In both cases the seasonal depth of frost penetration increases.

The behavior of the penetrating depth of thawing in the arctic mountains and zones is symmetrical. This means that with a decreasing absolute altitude the sizes of patterned grounds there increase in a corresponding way as from higher to lower latitudes of permafrost areas.

3. Aridity As a Favoring Factor for the Periglacial Environment, i.e., the Existence (Currency) of Patterned Grounds

Aridity can be understood as a favoring factor for patterned grounds in a double respect:

1) owing to the fact that patterned grounds at least can be met in many places of water-unsaturated debris covers and there are better developed than in water-saturated samples (Hövermann & Kuhle 1978). For this pleads the much clearer development in the limestone debris against the neighboring patterned grounds in the massif-crystalline - or phyllite-debris, where a significantly higher degree of water-saturation exists. This finding in the field has been proved by comparative investigations in the laboratory. In the temperature control chamber (Type Z 1200 e/20 Ju-S Weiss-system) debris of limestone- and granite -hornfels in styropor vats has been exposed once a day over c. seven years to the freezing- and thawing climate between +17° and -13°C registered in the Zagros-mountains (Kuh-i-Jupar: 39°N 57°E) at 3000 m asl. Result: the limestone-debris-cover developed faster and at last also clearer patterned grounds with a good sorting of material. The patterned grounds synthesized in the laboratory showed similarities with those in the limestone debris of the Kuh-i-Jupar (Zagros; Photo 13). When in a second test-row the precipitation had been lowered from 200 mm/a to 100 mm/a, the limestone remained inactive while in the granite-hornfels a forced development of patterned ground soon became significant. Generally at the same amount of rain in the granite-hornfels debris a four-times greater amount of water than in the limestone-debris has been measured. This proves the above-mentioned tendency to the better development of patterned grounds in water-non-saturated samples. That, however, only applies up to a certain degree of aridity. If this is exceeded, i.e., it becomes still drier, the development of patterned grounds is weakened. This is also proved by its intensifying in the granite-hornfels-debris in contrast to the samples of limestone at a decreasing precipitation. At the same time there can be summarized

that with regard to the optimum formation of patterned ground every rock, i.e., detritus type has its own specific moistness (Kuhle 1985c).

2) The second aspect that causes the aridity to become a favorite factor for the development of patterned grounds, at the same time concerns the spatial extension of the whole periglacial milieu where patterned grounds occur. It consists in the fact that the whole potential extension of the periglacial space is realized also empirically, i.e., effectively. Under humid conditions, however, part of it is covered by glaciers, so that there – below the glacier ice – no patterned grounds can occur. For instance, below the altitudinal level of the glaciers in the European Alps there exists only a relatively narrow permafrost belt of a few hundred meters in width (Figure 2) (Furrer & Fitze 1970). On it and then also further down – naturally as far as to the lower limit of patterned grounds - there exist patterned grounds, exactly like in arid mountains.

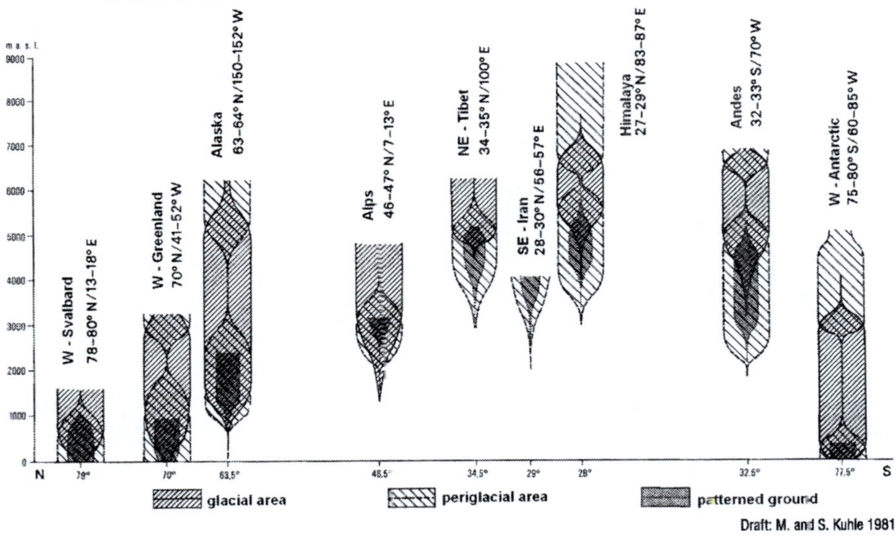

Figure 2. Nine empirical combinations, i.e., 7 types of combinations of glacial and periglacial altitudinal levels (areas, belts) in mountains of the earth.

In the European Alps this limitation due to the glacier cover lets become recognizable nearly no decrease in size of patterned grounds in the active layer towards above. The vertical distance of both altitudinal limits is too insignificant, i.e., the lower margin of the glacier is situated too close to the lower limit of the permafrost.

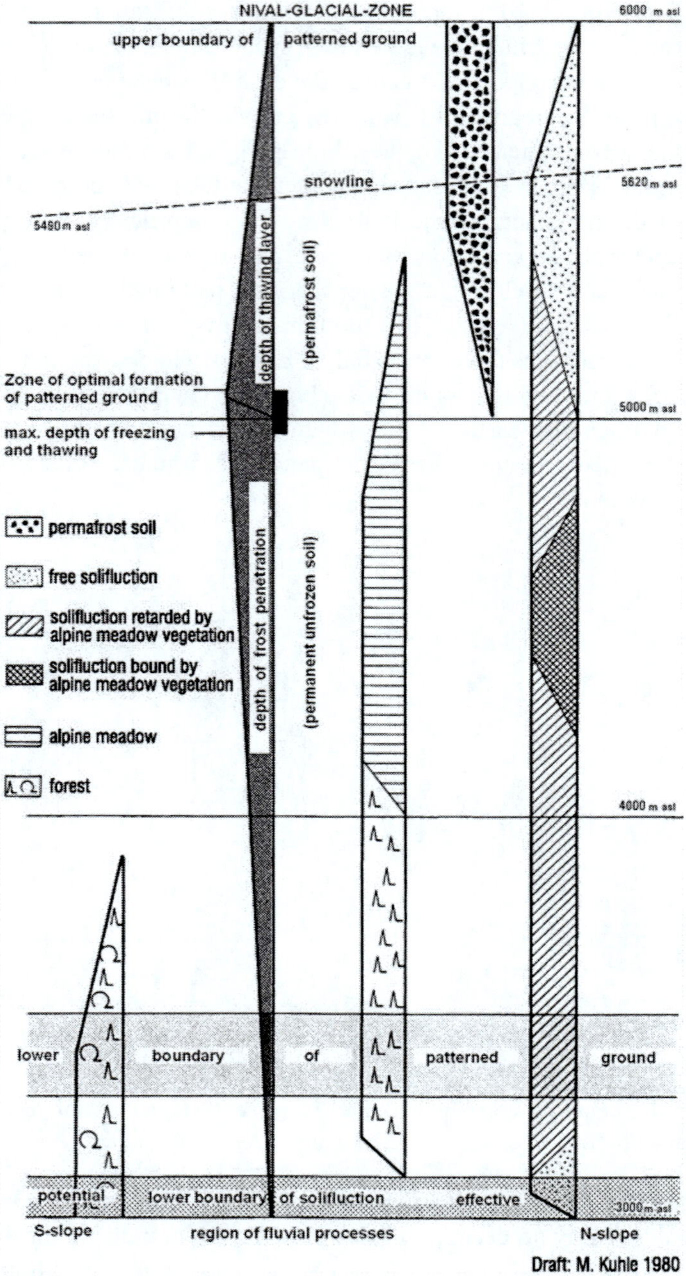

Figure 3. Periglacial area of the semi-arid Inner (Tibetan) Himalaya; example north of the Annapurna- and Dhaulagiri-groups (28°20'-28°55'N/83°-84°E) (cf. Figure 4 and 5).

In arid mountains like the subtropic Andes or the Tibetan Himalaya that in the precipitation shadow of the High Himalayas has to be described as at least semiarid, the decrease in size of the patterned grounds is evident (Figures 2, 3). In the Aconcagua massif (32 – 33°S) patterned grounds and related forms of frost change have been mapped downwards as far as at least 2600 m asl (Kuhle 1985b). The greatest patterned grounds – real macro-forms – have been met between c. 4100 and 4550 m asl (Photo 3). Above this zone of optimum development of patterned grounds around 4400 m asl, according to the height these have become smaller and smaller and at c. 5100 m asl have reached an upper boundary of patterned ground (Figure 2). Obviously this is a climatic upper boundary of patterned ground, because in places the debris slopes are stretching clearly higher without being covered by glaciers. However, they do not show patterned grounds anymore.

In the Inner– or Tibetan Himalayas solifluction occurrences (Kuhle & Jacobsen 1988: 598; Kuhle 2012: 194) and -forms as well as patterned grounds have been observed as far as below of 3500 m asl and – according to the developed depth of the valley – even down to c. 3000 m asl (Kuhle 1978a, c 1982a). Its size increases up to a height of c. 4800 m asl. For instance on the valley bottom of the Hidden Valley N of Dhaulagiri (28-30°N) there have been found patterned grounds from block circles of 2-3 m in diameter in this height (ibid.: Abb. 72) (Figure 1) (see also Fujii & Higushi 1976).

These are forms in arctic dimensions similar to those that can be observed in the Scandes (Jotunheim, Norway) on the Juvvasshö on the N-side of Galdhöpiggen at 1850 m asl (Photo 2) or at 1500 m asl in the massif of Kebnekaise (Lapland) north of the Arctic Circle. Boulders with an extension of the longitudinal axis of over 1 m are included into the circle-sorting. A further reference to the fact that this area is that one of the optimum development of patterned grounds, are stone- or boulder stripes in similar dimensions (Figure 1 (cf. Falkenhausen 1969)). Stone stripes are mainly developed on flatly inclined slopes and faces of mountain passes (e.g. on the Dhampus La west of the Thak Khola: Figure 1) situated at a greater distance from the talweg. Further above at 5200 m asl the solifluction forms and patterned grounds – independent of the condition of the rock-debris – already are clearly smaller. In addition, the slopes with an amorphous solifluction show a less clear alignment of longitudinal axes in the line of dip of the slope than at a minor altitude. From this one can also recognize the decrease of the forming tendency with increasing height above an optimum zone of patterned grounds. Here in the Tibetan Himalaya broad ridges of debris are developed up to 5800 m asl without a névé- and ice cover; in places they are situated even somewhat higher. The decrease in forming across a debris-ridge like this upwards, reaches between 5500 and 5700 m altitude – and locally still somewhat higher – an upper limit of forms of solifluction and patterned grounds. Due to the aridity this is no limit dictated by a névé- and ice-cover as it occurs in the humid mountains like the European Alps (see above; Figure 2). Here above, in the dry Tibetan Himalaya, as well as in many places of Tibet, an area is concerned on the surface of which debris for a frost-dynamic sorting is prepared very well; however, a sorting like this no longer takes place. Obviously this is the transition into a pergelide altitude area. These observations can be checked among others by climbing up to the 6012 m high Dhampus Peak (Marphatse) across the W-crest (ibid.: Abb. 12 No. 62) (Figure 1).

In the Arctic, several hundred kilometers away from the pole, similar observations of a decrease in forming towards above can be made – that is due to a relatively high ELA (Figure 1 W-Svalbard). Here, too, a certain aridity is proved. In the European Alps, however, the ELA is running absolutely nearly 2000 m higher, but in relation to the periglacial altitude level when compared with Spitsbergen clearly deeper. Accordingly, it is clearly more humid in the Alps (cf. Figure 2) (compare Fliri 1974 with Mortensen 1928). Thus in the Alps a decrease in size of patterned grounds above the optimum zone of development has nowhere been observed by the author.

4. THE OPTIMUM ZONE OF PATTERNED GROUNDS AS INDICATOR OF THE PERMAFROST LINE

In Spitsbergen the lower limit of patterned grounds is dictated by the sea-level (Figure 2 W-Svalbard) and in the climate-genetic sense would reach down even much deeper. This is already proved by the fact that there the permafrost reaches at least down up to the sea-level (Kuhle 1983a; cf. Köppen 1923). Accordingly, here the patterned grounds are soils of freezing and thawing induced by the thawing-process, whilst the area of patterned grounds, induced by the freezing process – that theoretically begins further down – still is not at all reached in Spitsbergen.

In the subtropical Andes and in the similarly semiarid marginal mountains of Tibet, i.e., in all places where the relief remains sufficiently far below under the permafrost line (Kuhle 1982b), the patterned grounds, induced by the process of freezing, develop with their diminishing size towards the lower margin of patterned grounds (Figure 2, 3 and 4).

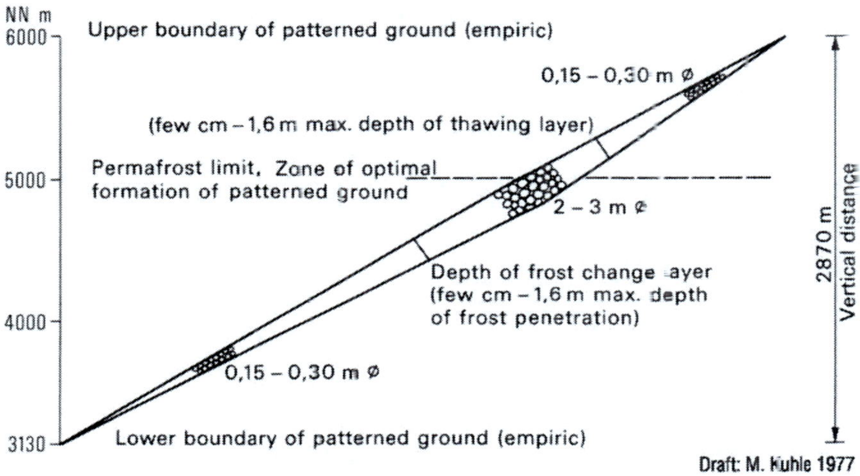

Figure 4. The approximately symmetrical periglacial altitudinal level (-area, -region) with the upper boundary of patterned ground, the optimal zone, -belt or -level of patterned ground and the lower boundary of patterned ground in the semi-arid Inner (Tibetan) Himalaya; example north of the Annapurna- and Dhaulagiri-groups (28°20'-28°55'N/83°-84°E) (cf. Figure 3 and 5).

Because the sizes, i.e., the diameters of patterned grounds, are directly connected to the depth of its sorting of material (c. 1.5:3; i.e., depth of sorting of material = 1.5 to diameter = 3 units), the patterned grounds grow larger with increasing depth of frost penetration. Because the depth of frost penetration increases with the absolute altitude, the patterned grounds, starting from their lower limit, grew larger and larger toward above (Figure 4).

Because the penetrating depth of thawing from the upper limit of patterned grounds toward below, i.e., with decreasing absolute altitude gets larger and larger, also the patterned grounds become larger and larger towards below (Figure 3 and 4).

Accordingly, patterned grounds reach their largest dimensions, where the thickness of freezing and thawing in the bottom is the greatest. Necessarily this is there, where the depth of frost-penetration is enough to create permafrost. This means that at the same time freezing and thawing is the greatest, where the penetration of thawing is the greatest.

It is exactly there, where the permafrost breaks off, i.e., at its lower limit. This is equivalent to the start of the permanent thawing soil. Accordingly, the soil of freezing and thawing reaches its greatest thickness where the permanent thawing soil and permafrost come into contact. This means, where the patterned grounds are the largest and, accordingly, have their greatest sorting depth, the permafrost-line is situated, that at the same time is the upper limit of the thawing soil (Figure 3, 4 and 5).

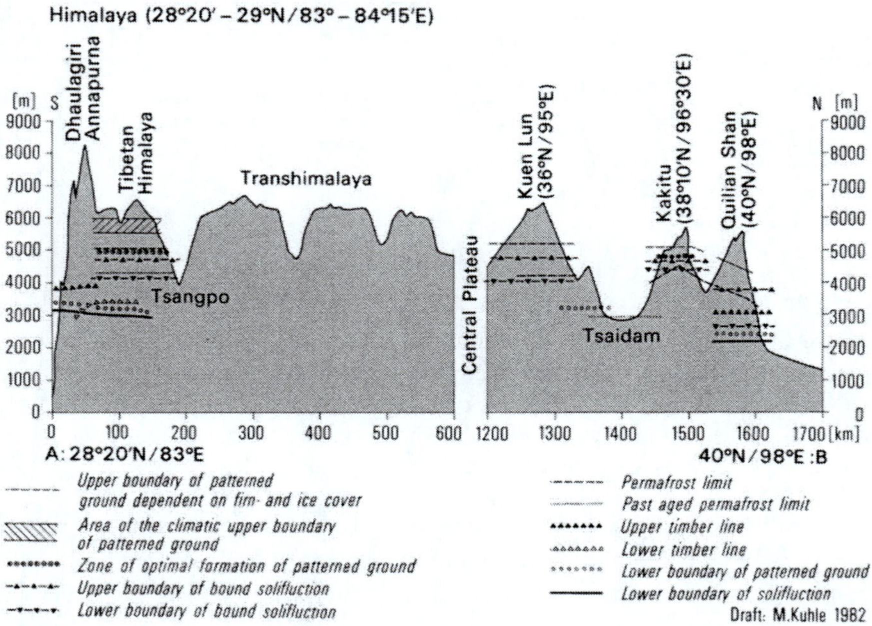

Figure 5. Levels and belts of the periglacial area (region) in Tibet between the High Himalaya and the Quilian Shan. Exaggerated c. 50 times.

Detailed observations in the field show that the upper limit of patterned grounds is dependent on the expositions. This can be explained by the dependence of the thickness of the thawing soil on the solar radiation. The lower limit of patterned grounds is controlled by the depth of frost-penetration, i.e., by freezing during the night and therefore (to a rather great extent, i.e., apart from the residual debris-warmth in the evening) independent on the exposition. With regard to this point the described symmetries between above and below of the permafrost-line are hurt.

5. An Arid (to Semi-Arid) and Thus Cold-Based Highland-Ice-Sheet Renders Sub-Glacial Permafrost Possible. This Is Proved by the Forms of Ground Moraines in Central Tibet

As has been shown above, the aridity enlarges the horizontal and vertical extension of the periglacial zone, i.e., altitude level. Here, the spreading of the given periglacial key forms as e.g. patterned grounds have been taken in consideration. These are surface forms that are stopped by the covering of glaciers. A characteristic of the complete periglacial zone is the permafrost. Occurrences of permafrost dovetail with the glacial zone and altitudinal level below the ELA between the tapering off glacier tongue termini (Photos 15 - 19). An overlapping of the two areas might be given at glacier-covered permafrost. It is an open question if permafrost is able to develop under a glacier cover. Due to its large thickness exceeding more than 1000 m, in Siberia it is seen as being fossilized and thus of glacial genesis. This interpretation implies a surface that was free of glaciers during glacial times or a decrease in temperature of such an extent at that time that the permafrost might have been developed even subglacially.

With regard to this important problem of an only subaerial or subglacial permafrost development, the author is not able to contribute any stable empirical results from Siberia. From Tibet, however, there exist several own findings that beyond the finding of actual permafrost (Photos 8 – 10; Figure 5) (cf. also An Zhongyuan 1980; Cheng Guodong 1982; Luo Xiangrui 1987; Wang Shaoling 1987; Zho Youwu & Guo Dongyin 1982) in fact plead for a subglacial development of permafrost. The past inland ice cover of Tibet has been proved by the covering with ground moraine (Kuhle 1985d; 2004, 2012; 2013) of at least 70% of the highland face (Figure 11; Photos 20 - 25). Further exemplary large-scale areas of ground moraine are situated between 4470 and 5200 m in height in Central Tibet, among others at 35°34'48"N 93°05'29"E, 4477 m asl; 36°02'46"N 88°36'19"E, 4900 m asl; 35°41'24"N 85°58'02"E, 5100 m asl; 33°31'47"N 89°02'32"E, 5175 m asl; 34°12'05"N 86°08'41"E, 5195 m asl; 33°32'19"N 89°02'00"E, 5200 m asl. The remaining 30% are interspersed by glacigenic erosion forms of glacier ground polishing in the bedrock (Photo 26) or they are flatly covered by late Late Glacial to Postglacial glaciofluvial gravel layers. These sandar-like loose rocks have been washed out of the moraines by meltwaters during and since the deglaciation of the ice sheet (Photo 19).

Photo 15. Central-Tibetan landscape of ground moraine S of the Tanggula Shan pass at c. 5190 m asl (in the surrounding of 32°54'30.74"N91°55'37.74"E). In the NW the 6010 m-massif of the S-Tanggula Shan can be seen. The ground moraine cover (■) is in the permafrost and in summer is superficially melted; it contains far-travelled granite boulders. Analogue photo M. Kuhle, 2/9/1989.

Photo 16. Central-Tibetan ground moraine landscape in permafrost (at c. 5020 m asl; 33°34'20.35"N 91°22'41.20"E) E of the 6621 m high Geladaindong massif (background). The ground moraine cover (■) shows the same condition as between the dead ice boulders in the area of the current glacier margins (cf. Photo 17, 19). As can be recognized by the traces in the active layer of the way (half-left above), currently its surface is slightly melted (half-left above). The clayey matrix (cf. Fig. 11) is situated between pebbly to head-sized components. Analogue photo M. Kuhle, 21/8/1989.

Photo 17. At c. 5450 m asl facing E via a permafrost area that is nearly completely covered with ground moraine (■), seen from the actual fore-field of the large Geladaindong-E-glacier as far as the horizon in Central Tibet (33°27'51.57"N 91°12'49.65"E). The frayed glacier margin (left margin) and the dead ice boulders are situated on the general level of the Central-Tibetan High Plateau about 5300 m asl. The ground moraine and the active layer are washed out near to the surface by glacial streams according to the season and reshaped. Analogue photo M. Kuhle, 22/8/1989.

Photo 18. Ground moraine covered with alpine meadow vegetation in permafrost at c. 5050 m asl in Central Tibet (33°33'5.78"N 91°21'26.74"E) in the N-Tanggula Shan taken facing W. The ground moraine (■) consists of a clayey-silty matrix (cf. Fig. 11) between up to block-sized components (cf. Photo 19) that are embedded isolated from each other; only along meltwater courses the pelites have been washed out of the ground moraine and residual gravel-bands as well as sander-faces were developed (□) Analogue photo M. Kuhle, 25/8/1989.

Photo 19. House-sized dead-ice-block (for comparison: small yak-caravan with loads of our equipment on the left) on still non-overgrown ground moraine in the permafrost at 5300 m asl (aneroid measurement) in Central-Tibet (33°28'17.51"N 91°13'16.43"E). Analogue photo M. Kuhle, 25/8/1989.

Photo 20. Permafrost area in the arid W-Central-Tibet (Aksai Chin) (34°32'43.50"N 80°25'12.15"E) at 5230 m asl (aneroid measurement). The vegetation-free ground moraine shows a superficial deflation-plaster of slab-debris. This corresponds with the outcropping sedimentary rocks in W-Tibet. Here, too, in the active layer the usual clay- and silt portions (cf. Fig. 11) of the moraine matrix – that by far are the largest part of the loose material – have been dug out. (▼) show tundra polygons (frost wedge or frost crack polygons), i.e., postglacial permafrost indicators, that could develop only after the deglaciation. Facing N in the background the Lungmu Co (lake). Analogue photo M. Kuhle, 9/9/1996.

Photo 21. Ground moraine cover in the permafrost area of E-Central-Tibet (c. 32°39'13.60"N 91°51'26.23"E) at 5010 m asl (aneroid measurement) between the N- and S-Tanggula Shan-pass. Since the late-Late-Glacial deglaciation, i.e., in the Holocene, the small brook has merely cut decimeter-deep into the active layer on the ground moraine surface and freed the polymictic erratic boulders (🠗) - from which part of the ground moraine has been built-up - from the fine material matrix. Among others these up to meter-sized boulders are far-travelled granite- and balsaltic boulders (🠗) that under the glacial inland ice must have been frozen to the ground moraine (see Section 5). Analogue photo M. Kuhle, 3/9/1989.

Photo 22. Example of a Central-Tibetan ground moraine cover in the permafrost-area, that in the area of its uppermost meter, i.e., its annual thawing bottom - that is the active layer - since the deglaciation has been reshaped by a brook, flowing down from S (on the right) to N (on the left) (🠗). Locality: 32°37'19.64"N 91°50'48.68"E, c. 5080 m asl. Here, meter-sized, round-edged to partly rounded far-travelled polymictic moraine boulders have been cleared of the pelitic matrix - into which they still were embedded below the ice – by fluvial erosion. The permafrost sheet (-table) below the active layer dams up the small pools and puddles; it hinders the seeping away of the monsoonal snowmelt-water. Photo taken facing ENE. Analogue photo M. Kuhle, 17/8/1989.

Photo 23. Permafrost ground with deflation plaster from superficially enriched ground moraine debris on a typical ground moraine cover in Central Tibet (c. 32°42'49.42"N 91°52'37.66"E, 5010 m asl). Embedded into the clay-containing pelitic matrix (cf. Fig. 11) of the ground moraine there can be seen polymictic, far-travelled, erratic boulders consisting of e.g. granite (light) and basalt (black). Due to the seasonal up-and down-movement of the permafrost sheet (permafrost table) the meter-sized boulders have been uplifted to the surface of the ground moraine cover and in many places are towering above this surface (🠗). Taken facing W. Analogue photo M. Kuhle, 3/9/1989.

Photo 24. Permafrost landscape with covering past ground moraine in the area of sedimentary rock that outcrops below, situated N of the northern Tanggula Shan pass in the surrounding of c. 33°6'33.53"N 91°52'49.28"E about 4900 m asl. (🠗) mark meter-sized far-travelled granite boulders "swimming" isolated from each other embedded in pelitic matrix (cf. Fig. 11). Currently there occurs fluvial linear erosion and in the active layer solifluction. On the hills, too, the ground moraine covers are showing traces of ground polishing of the Tibetan ice sheet (◓), as if they had been hardened under the cold-based glacier ice by subglacial permafrost. Analogue photo M. Kuhle, 17/8/1989.

Photo 25. Permafrost landscape in Central Tibet. Here in the ground moraine the summer active layer is c. 1 m thick. The conspicuous accumulation of up to several meter-sized boulders reaching some decimeters (the ice axt is 80 cm long) above the pelitic matrix of the ground moraine surface has to be led back to the frost heave in the active layer. The fact that several boulders are tilted over steeply speaks in favor of frost lifting (🠗). The rounded hills in the background are roches moutonnées and the larger ones have to be classified as glacial-streamlined hills. Traces of polishing in the mantling of ground moraine might be a hint to glacigenic ground polishing under subglacial conditions of permafrost. Photo facing W in the surrounding of c. 31°52'17.04"N 91°30'53.85"E, 4800 m asl (aneroid measurement). Analogue photo M. Kuhle, 4/9/1989.

Photo 26. Permafrost landscape in SW-Central Tibet (32°15'40.48"N 83°52'1.00"E) at 4400 (4256 aneroid-measurement) to 4900 m asl; taken facing S. The glacial (LGP = Last Glacial Period = MIS3) glacier cover has nearly completely mantled the mountain relief with ground moraine (■0) (cf. Fig. 11 No. 1 u. 2). A new and more insignificant Late Glacial advance of the glacier margin from the N, from the centre of Tibet, has run up the mountain slope and reshaped the glacial ground moraine cover with further ground moraine (■I). Both the moraine deposits must have happened during the time of an existing active layer. Obviously this has existed in a similar way during the early High Glacial as during the Late Glacial. Analogue photo M. Kuhle, 30/8/1996.

There are several scientists that are inconsistent to the author's glacial inland glaciation of Tibet (Burbank & Kang 1991, Shi et al. 1991, 1992, Hövermann & Lehmkuhl 1993, Owen & Benn 2005, Heymann et al. 2010). None of these authors, however, has contradicted the glacio-geological and – geomorphological findings of the author as e.g. large far-travelled, erratic boulders, ground moraines and their laboratory analyses, roches moutonnées, glacier polishings, trough valleys, endmoraines surrounding Tibet and their sediment-analyses in deep absolute altitudes up to below of 1000 m asl as well as paraglacial gravels preserved many hundred kilometers above the valley bottoms. On the contrary these many findings, collected since 1980 and distributed over the whole of High Asia (cf. Kuhle 2004 and 2013) have been ignored by these authors. Up to now also alternative explanations of origin as to the wealth of these glaciological key forms are completely lacking. Accordingly, these authors also did not try to prove a different genesis than the glacial genesis evidenced by the present author.

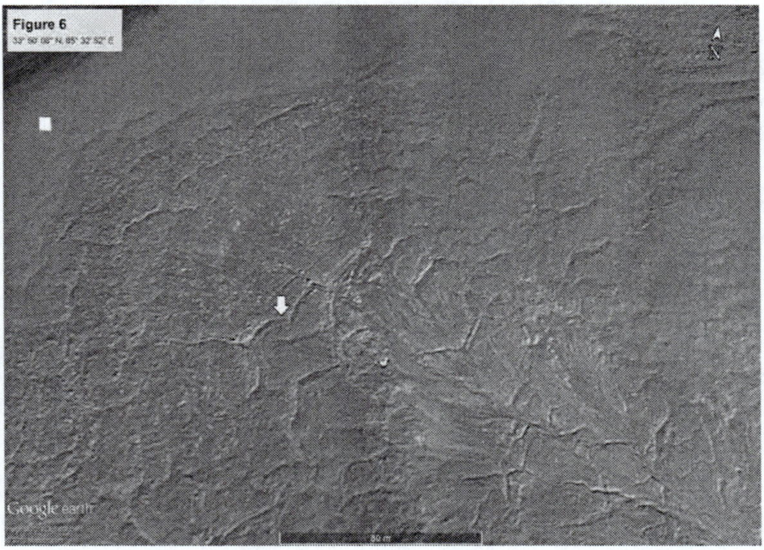

Figure 6. Satellite image: Ground moraine cover (■) under the influence of permafrost in Central Tibet about c. 5060 m asl. There are meter-sized, probably erratic boulders in the ground moraine. The ground moraine cover is – at least near to the surface – in a probably solifluidal movement to the ESE. (🡇) marks frost cracks belonging to a polygonal ice-wedge-network that probably is in the process of dissolution.

The exemplary section of ground moraine in Figure 6 shows moraine boulders in size of meters in a fine ground mass (matrix; cf. Figure 11), transported here. The ground moraine cover is passed through by landslip fissures and frost-cracks. The last belong to the incompletely developed phenomenoms of the community of forms of permafrost similar to networks of ice wedges (Tundrenpolygone). They have been developed since the deglaciation. The same is true for the completely developed ice-wedge-polygons in a ground moraine of the same sort in Figure 7. In the background this ground moraine cover is increasingly polished from W to E. Here, the ground polishing of the inland ice has left behind a lineation of exaration rills. The straight, parallel arrangement of the exaration rills over a distance in excess of kilometers to decakilometers, proves the large-scale corresponding direction of movement of a thick ice cover. This is the polish picture of a highly viscous to nearly rigid ice body (compare herewith Pillewitzer 1958, 1969). It points to an overlay of cold-based glacier ice.

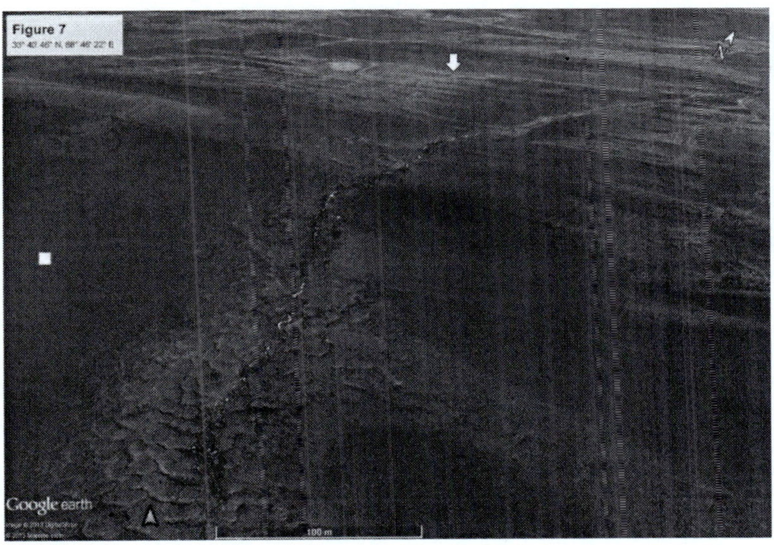

Figure 7. Satellite image of format-filling ground moraine: influence of subaerial (current) and subglacial (glacial) permafrost on the ground moraine cover (■) in Central Tibet about c.5180 m asl. In the ground moraine (■) are meter-sized, probably erratic boulders. From the left fore- to the middleground there stretches a 100 up to 200 m wide network of tundra-polygons (nets of ice wedges) (▲) that has been developed since the late Late Glacial deglaciation. This Holocene net of ice wedges reshapes, i.e., breaks through from below the W/to E running glacier scratches (⬇) of the inland ice in the ground moraine surface.

Photo 27. Ground moraine cover with boulders the size of meters in Central Tibet NE of Geladaindong (in the surrounding of 33°45'41.42"N 91°55'5.62"E; c. 4800 m asl). The hills in the background, polished like roches-moutonnèes, at least partly are covered with ground moraine. Possibly this has received its reworking by glacigenic ground polishing under subglacial permafrost conditions. Analogue photo M. Kuhle, 2/9/1989.

Photo 28. Permafrostlandscape with active layer in the SW of Central Tibet (31°20'36.12"N80°36'4.88"E i.e., 31°21'N 80°36'E; 4615 (aneroid measurement) – 4780 i.e., 4840 m asl); for subglacial permafrost as condition of the form of ground moraine (Fig. 11 No. 7) that mantles the bedrock in waves, there pleads its polished surface (striation: ▲). Under current subaerial conditions of a seasonal active layer, the ground moraine diagonally to the scratching is cut by fine microfluviatil rills (Mikrofluviatilrinnen) as far as down to the permafrost table situated at an insignificant depth (1-2 m). Analogue photo M. Kuhle, 2/9/1996.

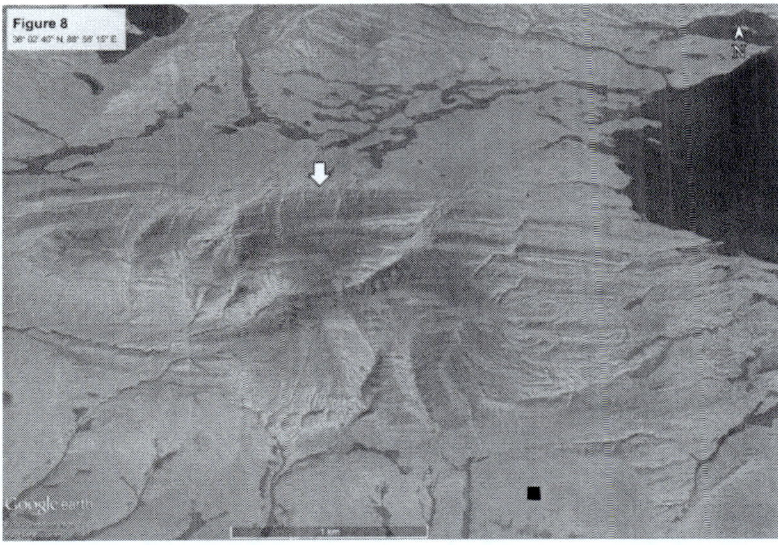

Figure 8. Satellite image of a ground moraine section: The subglacial influence of permafrost on the form of the ground moraine cover (■) in Central Tibet (c. 5030 m asl) can be made probable. i.e., recognized by the glacier scratches (⬇) in their surface. The lineation of the glacier scratches running from W to E on this bi-convex roche-moutonnée-like ground moraine ridge needs a hard surface for its development, i.e., subglacial permafrost without active layer. The actual subaerial permafrost, however, makes possible a down-slope forming of microfluviatil rills (Mikrofluviatilrinnen) that thus is running diagonally to the scratches in the active layer (on the right and left below ⬇). During the Holocene as far as today this cuts the past glacier scratches (cf. Photo 28).

By means of the decameter-thick ground moraines there can be established an analogy to outcropping bedrock, that astonishingly still today can be proved by its geomorphological condition. So, scratched roches moutonnées of moraine material are preserved (Figure 8). The precondition of their development is hard-frozen ground moraine that behaves like hard rock (cf. Photos 24, 25, 27, 28). This finding of subglacial permafrost is also confirmed, where similar glacier scratches are preserved less sharp on the edges. Figure 9 shows a state of preservation of this kind. The blunt striation is caused by the Postglacial, i.e., Holocene thawing soil, the active layer. Since the deglaciation it has allowed a seasonal solifluidal reshaping of the scratches.

In many places stream-line-shaped, round-polished moraine ridges have been observed (Photos 24, 27). Currently most of them are cut by small fluvial rills down-slope (Photo 28).

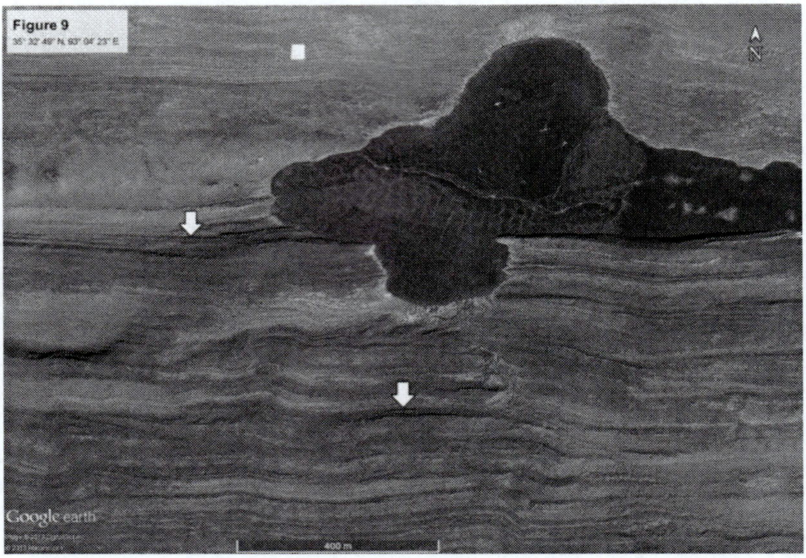

Figure 9. Satellite image: In this format-filling ground moraine cover, too, (■) there probably have been developed glacigenic scratches (🡇) under the hardening influence of subglacial permafrost and the ground polishing of the inland ice. They follow the large-scale incline and glacial discharge of the inland ice from W to E. The ground moraine section of Central Tibet chosen here is situated about 4495 m asl and thus at a 500 to 700 m minor height than the sections of Figures 7 and 8. Accordingly, the active layer is clearly thicker, what must have strengthened the reshaping since the deglaciation. Here, solifluction is concerned that in the meantime has made the lineation indistinct by scratches and exaration rills (🡇).

A further indicator of subglacial permafrost is shown in Figure 10. In its section a W-E-striated, many meter- if not decameter-thick ground moraine has been broken on a large scale into separate segments. The segments that due to permafrost remained compact are showing extensions of up to several 100 m. Frozen to the cold-based ice they obviously have been pulled out from their formation and slightly dislocated. Differently formed, partly wedge-shaped lateral faults, separated of the corresponding horizontal faults, are to be observed. According to the authors knowledge this phenomenon has not been observed in other inland ice areas. The mechanism in detail remains unknown. It might have been a High-Glacial lineation and late Late-Glacial subglacial crushing due to the Late Glacial uplift of the ELA – for instance by warming-up under a now thinner ice cover. At least a two-phase development like this is probable.

This is not at all an isolated case. A phenomenon like this has been observed at several localities in Central Tibet. So also at 34°12'05"N 86°08'41"E, 5195 m asl; 33°32'19"N 89°02'00"E, 5200 m asl.

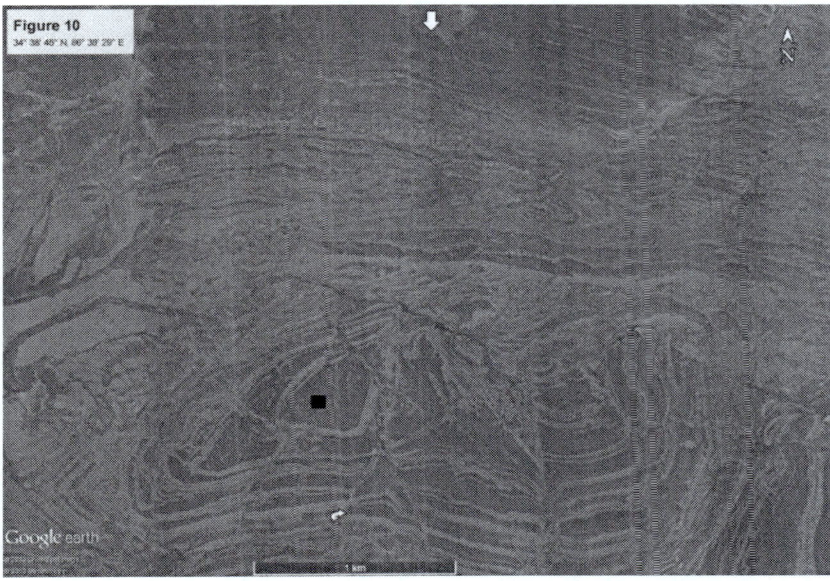

Figure 10. Satellite image of a further Central-Tibetan section in the field with an exemplary covering of ground moraine (■) under the influence of permafrost. The surface is situated about 5050 m asl and shows a W/E-directed lineation i.e., scratching (🡇). In places this lineation or scratching is broken and has been shifted against each other by lateral faults i.e., along of horizontal transcurrents ('Blattverschiebungen') (↻). Faults or joins like these encircle several hundred meters- up to more than 1 km - extended clods (↻). Presumably they were frozen to the ground of the cold-based inland-ice and due to glacier movement have been torn out of their original formation. The partly compact preservation of the clods itself points to their stability under the influence of permafrost. Primary lineation and secondary breaking with dislocations point to a development of these forms in two phases.

5.1. Climate-Genetic Possibility to Explain Subglacial Permafrost in the Glacial Tibet

During glacial times the high plateau of Tibet nearly completely belonged to the glacier feeding area (Kuhle 1987a). The ablation area of the inland ice was its margin, falling away steeply and deeply into the subtropical and monsoon-tropical climate.

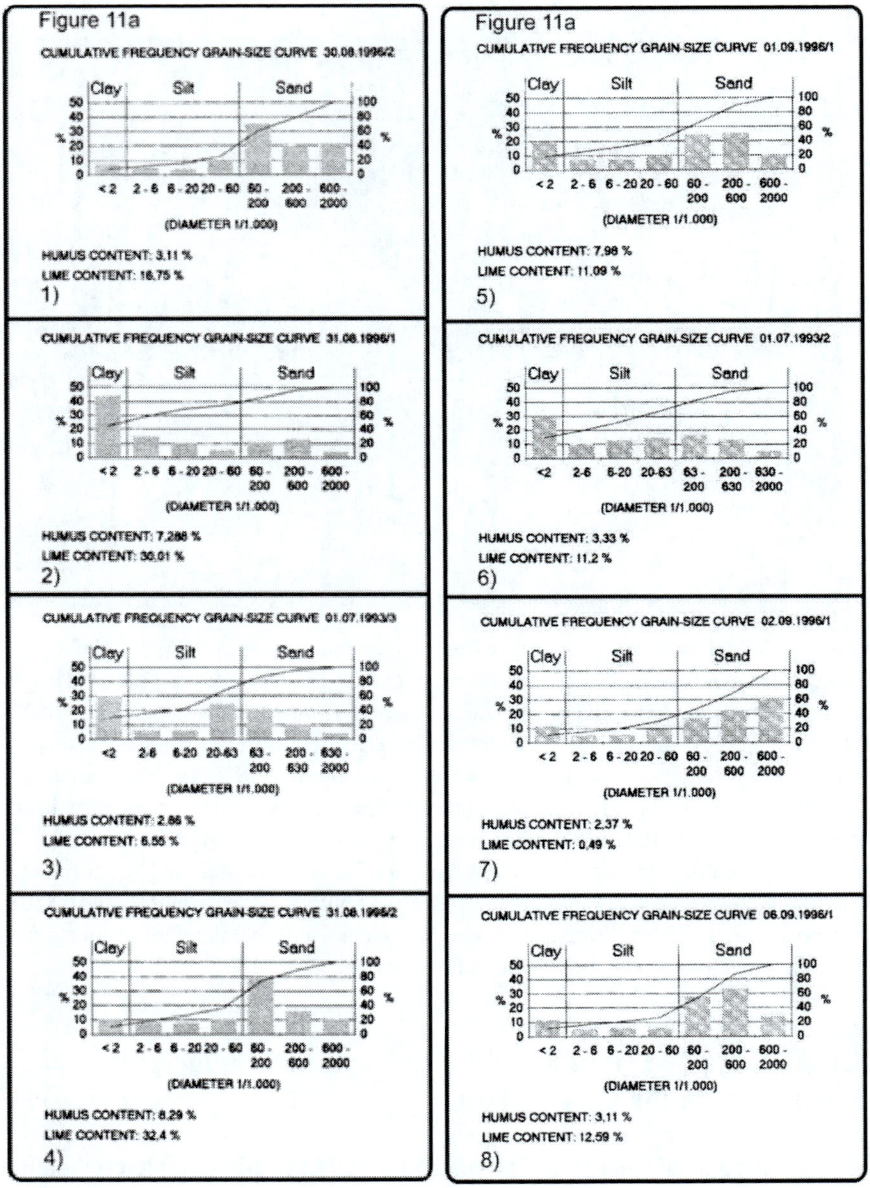

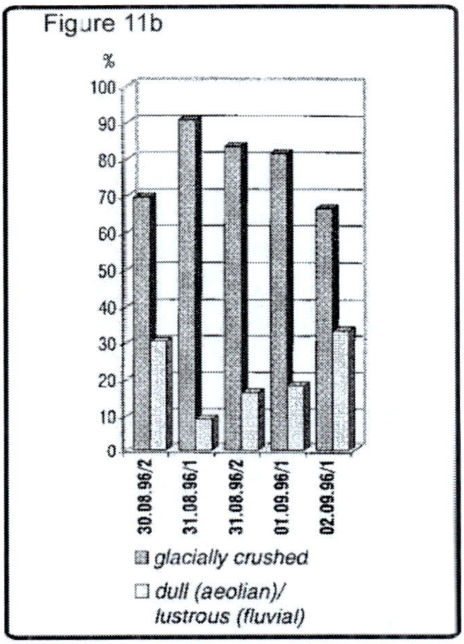

Probennr./Datum; sample No/date	*0,2 - 0,6 mm ausgezählte Quarzkörner*; 0,2 - 0,6 mm counted quartz grains	*glazigen-gebrochen/ frisch (in situ) verwittert*; glacially crushed/ freshly weathered (in situ)	*äolisch mattiert*; dull (aeolian)	*fluvial poliert*; lustrous (fluvially polished)	*Anmerkungen*; remarks
30.08.96/ 2	158	69,6%	19,0%	11,4%	*ca. 50% Quarz*; c. 50% quartz
31.08.96/ 1	88	90,9%	9,1%	-	*nur ca. 20% Quarzanteil, sehr kantig-frisches Bruchmaterial, teils oberflächlich modifiziert*; only c. 20% quartz portion, well edged fresh fracture material, partly surficially modified
31.08.96/ 2	91	83,5%	14,3%	2,2%	*frisch gebrochener Detritus*; freshly broken detritus
01.09.96/ 1	206	81,6%	13,8%	4,6%	*klassifiziertes, frisch gebrochenes Substrat, teilweise leicht fluvial überarbeitet*; classified freshly broken substrate, in parts slightly fluvially reworked
02.09.96/ 1	102	66,6%	6,9%	26,5%	-

Figure 11. 6 representative samples of ground moraine matrix from the Central-Tibetan permafrost-area of the past Tibetan Ice Sheet and 2 comparative samples from the ground moraine- and permafrost area of the Laurentide Ice Sheet in Canada aside of morphoscopic analyses of the Tibet-samples: sampling of all 8 samples M. Kuhle.

1) Sample locality: E of the locality of Photo 26 at 32°5'23.89"N 84°53'52.92"E; 4440 (4365 aneroid-measurement) m asl, depth of sampling 0.15 m; characteristics of the moraine: tri-modal course of the cumulative-curve; clay peak as Tertiary maximum. See morphoscopy.

2) Sample locality: c. 32°31'N82°30'E, 4450 (4270 aneroid measurement at high pressure weather situation) m asl. Locality: W of Photo 26; sampling-depth 0.15m; limestones are outcropping in the immediate catchment area, so that the lime content points to a 30% portion of local moraine on the ground moraine. Characteristics of the moraine: fine-corn-peak with 43% clay that predominates the other fractions. For comparison see

3) ground moraine sample from Canada. There, too, the clay dominates the pelite-portions though the limestone-content contains only c. 1/5 of that one of the sample from Central-Tibet. Accordingly, the high clay portion has to be understood as an immediate clue to the strong trituration by a thick overlay of inland ice. See morphoscopy. 3) Sample for comparison of 2) Sample locality: West Canada 50°17'41.26"N 107°55'10.83"W, 750 m asl; 4 km W of Swift Current 40 km S of the South Saskatchewan River. Matrix from Last Glacial Maximum (LGM) ground moraine of the Laurentide Ice Sheet deposited in the past permafrost area.

4) Sampling locality c. 32°4'49,78"N 81°59'58.19"E, 4850 (4675 aneroid measurement) m asl. High- to Early-Late Glacial limestone-containing matrix from an area where among others limestone outcrops in the underground; sampling depth 0.2 m. Characteristics of moraine e.g.: the matrix contains far-travelled, isolated erratic components; see morphoscopy: c. 82% glacially crushed SiO^2 grains.

5) Sampling locality: c. 32°31'20"N 80°45'10"E; 4500 (4325 aneroid measurement) m asl, at the N- source-branch of the Indus-valley, here named Senko Tsangpu, SW-Central-Tibet; sampling-depth: 0.5 m; characteristics of the moraine: though locally there are outcropping mainly lime stones, the limestone portion is relatively insignificant, what confirms the substrate as being far-travelled moraine from the center of W-Tibet. The important clay-portion (c. 19%) is characteristic of an inland-ice-typical, strongly triturated kind of detritus, grinded by the heavy weight of the ice burden. See similarity with comparison-sample from Canada (6)). Cf. morphoscopy.

6) Comparison sample to 5) sample-locality: Canada c. 50°26'26.90"N 105°34'40.16"W, 580 m asl, NW of Moose Jaw c. 100 m ENE of Lake Diefenbaker (W-section of the Late Wisconsin Laurentide Ice Sheet).

7) Sample locality: S-Tibetan permafrost-landscape with active layer 31°21'N 80°36'E, Photo 28 in the foreground; 4780 (4615 aneroid-measurement) m asl; sampling depth: 0.15 m; moraine characteristics: perfect fine grain peak in the clay

(11%) as well as erratic granite boulders, embedded isolated from each other in the matrix. See morphoscopy.

8) Sample-locality: Southwestern Tibet-plateau, Gar Zangbo (upper Indus valley) c. 31°38'30"N 80°21'E, NW of Photo 28; 4510 (4365 aneroid-measurement) m asl; sampling depth: 0.2 m. Characteristics of the glacigenic sediment: a clay-portion of 11% is typical of ground moraine; also the bimodal course of the curve of the heights of the columns; the important medium- and fine sand portions (c. 32% and 28%) are characteristic of the – in comparison with the ground moraine- coarser ablation moraine as far as to the 'drift-cover-sand' ('Geschiebedecksand') developed from them. Morphoscopy: 60.9% = glacially crushed, 11.7% = dull (aeolian), 27.4% = lustrous (fluvially polished).

Therefore it was – in comparison with the ablation areas of the flatland ices – small and nearly the entire outline of the Tibetan highland was situated above the snowline (ELA) (Kuhle 1997; 2012). The monsoon, collapsed because of the stabile cold high of the Tibetan ice (Kuhle 2001) caused an arid, albedo-intensive highland ice, that due to an extreme evaporation-coldness in addition was super-cooled. A subtropical and therefore very high-lying inland ice like this has no parallel – not even in Siberia or from Alaska via Canada up to Greenland. A comparison of the European inland ice with the Tibet ice has to be ruled out anyway, because it was a wet-based inland ice.

Thus perhaps subglacial permafrost even is an exclusive characteristic of an arid and hypsometrically super-cooled highland-inland-ice as it has existed in Tibet **alone**.

As to current glaciological analogies the Antarctic- and Greenland-inland ices are appropriate. According to the definition of Shumskii (1947, 1964: 407, 408 and Figure 122; s. also Müller 1962) they both are situated with the centers of their nourishing areas in the area of the recrystallization zone. Today in the Himalaya this nourishing-area – i. e. the recrystallization zone – is reached at an altitude in excess of 7000-7100 m asl (Kuhle 1988: 550). This is an altitudinal level where the temperatures do not overstep the freezing-point of -0°C throughout the year (Kuhle 1986). If during the Last Glacial Period (LGP) this altitudinal level was lowered by c. 1200 m – what would correspond with the glacial ELA-depression (Δ ELA) – then it had fallen down large-scale under the surface height of the glacial Tibet-Ice. This could be an additional pre-condition of subglacial permafrost.

6. Current Global Combinations of Periglacial Areas (Levels) with Patterned Grounds and Glacial Areas (Levels) (Figure 2)

Empirical investigations showed, that the planetary change of forms is repeated in the hypsometric change of forms. Accordingly, the periglacial area is repeated from the areas close to the pole to the areas far away in the increasing periglacial altitudinal levels of the mountains (Figure 2). Due to the aridity there that reduces the covering of the glaciers (see Sections 3 and 4), it does not reach its greatest heights in the tropics, but in the subtropics. Therefore a sensible description of the periglacial area is not possible without the registration of its clear dividing line to the glacial area (cf. Köppen 1923). The periglacial area is situated always climatically and in many places also spatially on the periphery of the glacial area. Therefore its ice cover in the direction of the poles is climatically comparable to that one due to the mountain glaciers. However, as to this empirical comparison the topographic characteristics of the glacial areas connected towards above must be considered. This is necessarily due to the interlocking of the two areas below of the snowline (ELA), because the ablation areas of the glaciers and thus of the glacial area, situated below the ELA, are reaching the deeper into the periglacial area, the higher the relief, i.e., the nourishing area of the glaciers towers above the ELA.

From this combination of climate and relief in the investigation areas that the author has met during his researches in the field, seven empirical types have been extracted, characterizing the periglacial area – and in it the patterned grounds – in its proportion to the glacial area.

Figure 2 sketches:

Type 1 according to the example W-Svalbard: The periglacial area reaches with patterned grounds and even with permafrost the sea-level; it is dovetailed with the glacial area, so that it tapers out towards above; a climatic patterned ground-upper limit has not been developed, but towards above the size of the patterned ground decreases.(Photos 1 and 14).

Type 2 according to the example W-Greenland: The periglacial area reaches with patterned grounds and permafrost nearly the sea-level, it is dovetailed with the glacial area, so that it tapers out towards above; the size of the patterned grounds does not decrease towards above, i.e., the patterned grounds reach the glacier margin approx. in the area of its optimum zone of development. – Type 2 is also realized in the W-Antarctic.

Type 3 according to the example Alaska: The periglacial area reaches with patterned grounds its lower limit in the permanent thawing soil, so that the patterned grounds increase in size towards above; the periglacial area and patterned grounds are dovetailed with the glacial area; up to the glacier margin the patterned grounds do not exceed the height of their optimum zone of development.

Type 4 according to the example of the European Alps: The glacial area breaks through the periglacial area towards below in the area of the ever-thawing- soil, limiting the patterned grounds towards above in a way, that the optimum development zone of patterned grounds, i.e., the permafrost limit is only just reached.

Type 5 according to the example of NE-Tibet: A large part of the periglacial area is not covered by the glacial area, so that an optimum development zone of patterned ground is realized (Photo 4), but the upper limit of patterned grounds is not quite reached.

Type 6 according to the sample of SE-Iran: The mountains in the Zagros show a periglacial area with patterned grounds from the lower limit of which the size of patterned grounds increases towards above (Photo 13) without reaching their optimum zone of development; the summits are reaching neither the permafrost limit nor the height of the glacial area.

Type 7 according to the sample of Himalaya: There is developed a lower limit of the periglacial area where the solifluction ceases, as well as a patterned-ground lower margin; the size of the patterned grounds reaches with its optimum zone of development the lower limit of permafrost and decreases above up to an empirical patterned-ground upper limit; this upper limit is only realized on debris-covered mountain-ridges in the precipitation shadow on the Himalaya-N-side; the periglacial area is dovetailed with the glacial area.- Type 7 is also realized in the subtropical Andes.

ACKNOWLEDGMENT

The author likes to thank Mrs Juliane Jörgens for the translation of this manuscript.

References

An Zhongyuan, 1980. Formation and evolution of permanent ice mound Qing-Zang Plateau. *J. Glaciology and Cryopedology*. 2 No. 2, Lanzhou, 25-30.

Ball, D. F.; Goodier, R., 1970. Morphology and distribution of features resulting from frost-action in Snowdonia. *Field Studies* 3, 193-217.

Boesch, H., 1951. Beitrag zur Kenntnis der Blockströme. *Die Alpen* 27, 1-5.

Brosche, K. U., 1971. Beobachtungen an rezenten Periglazialerscheinungen in einigen Hochgebirgen der Iberischen Halbinsel. *Die Erde* 102/1, 34-52.

Büdel, J., 1959. Periodische und Episodische Solifluktion im Rahmen der klimatischen Solifluktionstypen. *Erdkunde* 13, 297-314.

Burbank, D. W., Kang Jiang Chen, 1991. Relative Dating of Quaternary Moraines, Rongbuk Valley, Mount Everest, Tibet: Implications for an Ice Sheet on the Tibetan Plateau. *Quarternary Research* 36, 1-18.

Cheng Guodong, 1982. The Geographic Laws of Permafrost Distribution in Qilian Shan. *Braunschweigische Wissenschaftliche Gesellschaft Sonderheft 6 (Chinesisch-deutsche Tibet-Expedition 1981, Braunschweig Symposium 14.-15.4.1982) Hersg.: K. H. Olsen &. U. Wannagat*, 28-29.

Cailleux, A.; Taylor, G., 1954. Cryopèdologie, ètude des sols gèles. *Act. Scient. et Industr*. 1203, 1-218.

Dukiewicz, L., 1968. Congelisolifluction Lobes on the South Coast of Hornsund, Vestspitsbergen. Polish Spitsbergen Expedition 1957-60. *Polish Academy of Science*, 259-264.

Falkenhausen, E.v., 1969: Ein Blockmeer im Himalaya. *Zeitschrift für Geomorphologie* N.F. 13, 506-511.

Fliri, F., 1974. Niederschlag und Temperatur im Alpenraum. *Wiss. Alpenvereinshefte*, Nr. 24, Innsbruck, 1-111.

Fränzle, O., 1959. Glaziale und periglaziale Formbildung im östlichen kastilischen Scheidegebiete (Zentralspanien). *Bonner Geographische Abhandlungen* 26, 1-80.

Fujii, Y., Higushi, K., 1976. Ground temperature and its relation to permafrost occurences in the Khumbu region and Hidden Valley. *Journal of the Japanese Society of Snow and Ice (Seppyo)* 38 (Special Issue), 125-129.

Furrer, G., 1965. Die subnivale Höhenstufe und ihre Untergrenze in den Bündner und Walliser Alpen. *Geographica Helvetica 20*, 185-192.

Furrer, G., 1986. Solifluidale Bodenformen und Gliederung der subnivalen Höhenstufe im Braldo- und Biafo Tal (Karakorum). In: *Internat. Symposium über Tibet u. Hochasien 1985;* Kuhle, M. (Ed.) *Göttinger Geogr. Abh.* 81, 143-144.

Furrer, G., Fitze, P., 1970. Beitrag zum Permafrostproblem in den Alpen. *Vereinsjahresschrift der Naturforschenden Gesellschaft Zürich* 115, 3, 353-368.

Graf, K., 1973. Vergleichende Betrachtungen zur Solifluktion in verschiedenen Breitenlagen. *Zeitschrift für Geomorphologie* N. F. 16, 104-154.

Grötzbach, E., 1965. Beobachtungen an Blockströmen im afganischen Hindukusch und in den Ostalpen. *Mitt. Geogr. Ges. München* L, 175-201.

Hastenrath, S., 1973. Observations on the periglacial morphology of Mts. Kenya and Kilimanjaro, East Africa. *Zeitschrift für Geomorphologie* N.F. 16, 161-179.

Heim, A., 1936. The glaciation and solifluction of Minya Gongkar. *Geographical Journal* 87, London, 444-450.

Heyman, J., Stroeven, A. P., Caffee, M. W., Hättestrand, C., Harbor, J., Li, Y., Alexanderson, H. Zhou, L., Hubbard, A., 2010. Palaeoglaciology of Bayan Har Shan, NE Tibetan Plateau: the case of a missing LGM expansion. In: Heyman, J., Palaeoglaciology of the northeastern Tibetan Plateau. *Dissertations from the Department of Physical Geography and Quaternary Geology, Stockholm University, Sweden*. No. 21. Paper V, 1-15.

Höllermann, P., 1967. Zur Verbreitung rezenter periglazialer Kleinformen in den Pyrenäen und Ostalpen. *Göttinger Geographische Abhandlungen* 40, 1-98.

Hövermann, J., 1972. Die periglaziale Region des Tibesti und ihr Verhältnis zu angrenzenden Formungsregionen. *Göttinger Geographische Abhandlungen* 60, 261-283.

Hövermann, J.; Kuhle, M., 1978. Verbreitung und Bildung von Frostmusterböden in den Gebirgen des Vorderen Orients und der Sahara. *Coll. sur le Périglaciaire d'Altitude du Domaine Mediterranée et abords*, IGU et CNFDG, Strassbourg, 321-332.

Hövermann, J.; Lehmkuhl, F., 1993. Bemerkungen zur eiszeitlichen Vergletscherung Tibets. *Mitteilungen der Geographischen Gesellschaft zu Lübeck* 58, 137-158.

Huang Rongfu, 1982. Vegetation of the periglacial Zone in the NE Part of the Quinghai-Xizang Plateau. *Braunschweigische Wissenschaftliche Gesellschaft Sonderheft 6 (Chinesisch-deutsche Tibet-Expedition 1981, Braunschweig Symposium 14.-15.4.1982) Hersg.: K. H. Olsen & U. Wannagat*, 50-51.

Kelletat, D., 1969. Vergesellschaftung rezenter Periglazialerscheinungen im Apenin. *Göttinger Geogr. Abh.*, 48, pp. 1-144.

Klaer, W., 1977. Zur Morphogenese und Formungsdynamik des periglazialen Stockwerkes der Hochgebirge Vorderasiens. In: *Formen, Formengesellschaften und Untergrenzen in den heutigen periglazialen Höhenstufen der Hochgebirge Europas und Afrikas zwischen Arktis und Äquator. Abhandlungen der Akademie der Wissenschaften Göttingen, Mathematisch-physikalische Klasse* Nr. 3 31 (Bericht über ein Symposium), 261-274.

Köppen, V., (1923): Die Klimate der Erde: Grundriss der Klimakunde. Walter de Gruyter & Company, 1-369.

Kuhle, M., 1978a. Obergrenze von Frostbodenerscheinungen. *Zeitschrift für Geomorphologie* N. F. 22 (3), 350-356.

Kuhle, M., 1978b. Über Solifluktion und Strukturböden in SE-Iranischen Hochgebirgen. *Builetyn Periglacjalny* 27, 117-131.

Kuhle, M., 1978c. Über Periglazialerscheinungen im Kuh-i-Jupar (SE-Iran) und im Dhaulagiri-Himalaya (Nepal) sowie zum Befund einer Solifluktionsobergrenze. In: *Coll. sur le Périglaciaire d'Altitude du Domaine Méditerranée et abords, IGU et CNFDG*; Straßburg: 289-310.

Kuhle, M., 1982a. Der Dhaulagiri- und Annapurna-Himalaya. Ein Beitrag zur Geomorphologie extremer Hochgebirge. *Zeitschrift für Geomorphologie Supplement* 41 (Suppl. Bd.), Bd. I (Text): 1-229; Bd. II (Abb. mit Geomorph. Karte 1:85 000): 1-184.

Kuhle, M., 1982b. Erste Chinesisch-Deutsche Gemeinschaftsexpedition nach NE-Tibet und in die Massive des Kuen-Lun-Gebirges 1981. Ein Expeditions- und vorläufiger Forschungsbericht. In: *Tagungsber. und wiss. Abhandlungen des 43. Deutschen Geographentages Mannheim 1981.* (Eds: Hagedorn, H.; Giessner, K.) Steiner, Wiesbaden, 63-82.

Kuhle, M., 1983a. Zur Geomorphologie von S-Dicksonland (W-Spitzbergen) mit Schwerpunkt auf der quartären Vergletscherungsgeschichte. *Polarforschung* 53 (1), 31-57.

Kuhle, M., 1983b. Postglacial Glacier Stades of Nugssuaq Peninsula, Westgreenland (70° 03'- 70° 10'N). In: *Colloquium Trier 15.-17.5.1980: Late- and Postglacial Oscillations of Glaciers: Glacial and Periglacial Forms.* Ed: Schroeder-Lanz, H. (In memoriam Hans Kinzl), 325-355.

Kuhle, M., 1983c. Permafrost and Periglacial Indicators on the Tibetan Plateau from the Himalayan Mountains on the South to the Quilian Shans on the North. In: *Permafrost, Fourth International Conference, Fairbanks,*

Alaska, U.S.A. July 18-22, 1983 Abstracts and Program. University of Alaska, Fairbanks, ISBN 0-915360-03-9, 159-160.

Kuhle, M., 1985a, Permafrost and Periglacial Indicators on the Tibetan Plateau from the Himalaya Mountains in the South to the Quilian Shan in the North (28-40°N). *Zeitschrift für Geomorphologie* N.F. 29 (Bd. 2), 183-192.

Kuhle, M., 1985b. Bilaterality of the Periglacial Zone in the Aconcagua Massif (32°-33°S/70°W). In: *Acta Periglaciales (Geocriogenica) 1 RA (1). Union Geographica International,* Argentinien (ed.), 18.

Kuhle, M., 1985c. Experimental Synthesis of Patterned Ground and Consequences for a Supra-regional Comparison of the Lower Boundaries of Patterned Ground. In: *Acta Periglaciales (Geocriogenica) 1 RA, 3 (85). Union Geographica International*, Argentinien (ed.), 19.

Kuhle, M., 1985d. Glaciation Research in the Himalayas: A New Ice Age Theory. In: *Universitas 27* (4), 281-294.

Kuhle, M., 1986. The Upper Limit of Glaciation in the Himalayas. *GeoJournal 13* (4), 331-346.

Kuhle, M., 1987a. Subtropical Mountain- and Highland-Glaciation as Ice Age Triggers and the Waning of the Glacial Periods in the Pleistocene. *GeoJournal 14 (4),* 393-421.

Kuhle, M., 1987b. Glacial, Nival and Periglacial Environments in Northeastern Quinghai-Xizang-Plateau. In: Hövermann, J., Wenjing, W. (eds.): *Reports on the Northeastern part of the Quinhai-Xizang (Tibet)-Plateau by the Sino-West-German Scientific Expedition 1981.* Science Press, Beijing, China, 176-244.

Kuhle, M., 1988. Topography as a Fundamental Element of Glacial Systems. A New Approach to ELA-Calculation and Typological Classification of Paleo- and Recent Glaciation. *GeoJournal 17 (4,* Eds: Kuhle, M., Wenjing, W.: *Tibet and High-Asia - Results of the Sino-German Joint Expeditions (I)),* 545-568.

Kuhle, M., 1997. New Findings concerning the Ice Age (Last Glacial Maximum) Glacier Cover of the East-Pamir, of the Nanga Parbat up to the Central Himalaya and of Tibet, as well as the Age of the Tibetan Inland Ice. *GeoJournal 42 (2-3,* Ed: Kuhle, M.: *Tibet and High Asia - Results of Investigations into High Mountain Geomorphology, Paleo-Glaciology and Climatology of the Pleistocene (Ice Age Research) IV,*, 87-257.

Kuhle, M., 2001. The Tibetan Ice Sheet; its Impact on the Palaeomonsoon and Relation to the Earth's Orbital Variations. *Polarforschung* 71 (1/2), 1-13.

Kuhle, M., 2004. The High Glacial (Last Ice Age and LGM) ice cover in High and Central Asia. *Development in Quaternary Science* 2c *(Quaternary Glaciation - Extent and Chronology, Part III: South America, Asia, Africa, Australia, Antarctica*, Eds: Ehlers, J.; Gibbard, P.L.), Elsevier B.V., Amsterdam, 175-199.

Kuhle, M., 2012. The Former Tibetan Ice Sheet. In: *Ice Sheets: Dynamics, Formation and Environmental Concerns.* Eds. J. Müller & L. Koch; Nova Science Publisher, Inc. New York, 173-203.

Kuhle, M., 2013. The uplift of High Asia above the snowline and its Glaciation as an albedo-dependent cause of the Quaternary ice ages. *Nova Science Publishers Inc.*, New York, 1-240.

Kuhle, M.; Jacobsen, J. P., 1988. On the Geoecology of Southern Tibet - Measurements of Climate Parameters including Surface- and Soil-Temperatures in Debris, Rock, Snow, Firn and Ice during the South Tibet- and Mt. Everest Expedition in 1984. *GeoJournal 17 (4,* Eds: Kuhle, M., Wenjing, W.: *Tibet and High-Asia - Results of the Sino-German Joint Expeditions (I))*, 597-615.

Kuhle, M.; Roesrath, C., 1990. Geographie und Geologie der Hochgebirge. *BLV Verlagsgesellschaft (Kommission für Alpin-Lehrpläne des Deutschen Alpenvereins)*, München/ Wien/ Zürich, ISBN 3-405-13604-0, 1-160.

Luo Xiangrui, 1987. Relationship between the Distribution of Permafrost and Vegetation in the Anyemaqen Mountain Region. In: Hövermann; J., Wenjing, W. (eds.): *Reports on the Northeastern part of the Quinhai-Xizang (Tibet)-Plateau by the Sino-West-German Scientific Expedition 1981*. Science Press, Beijing, China, 367-386.

Louis, H.; Fischer, K., 1979. Allgemeine Geomorphologie. 4. Auflag, Walter de Gruyter & Co. Berlin, 1-815.

Müller, F., 1962. Zonation in the accumulation area of the glaciers of Axel Heiberg Island. N.T.W., Canada. *Journal of Glaciology* 4, 33, 203-210.

Mortensen, H., 1928. Über die klimatischen Verhältnisse des Eisfjord-Gebietes. *Chemie d. Erde* 3 (3-4), 611-638.

Ohlson, B., 1964. Frostaktivität, Verwitterung und Bodenbildung in den Fjeldgegenden von Enontekiö, Finnisch-Lappland. *Publ.Inst.Geogr.Univ. Turku* 37, 1-180.

Owen, L. A., Benn, D. J., 2005. Equilibrium-line altitudes of the Last Glacial Maximum for the Himalaya and Tibet: an assessment and evaluation of results. *Quaternary International* 138-139, 55-78.

Pillewitzer, W., 1958. Neue Erkenntnisse über die Blockbewegung der Gletscher. *IAHS AISH Publication* 46, 429-436.

Pillewitzer, W., 1969. Die Bewegung der Gletscher und ihre Wirkung auf den Untergrund. *Zeitschrift für Geomorphologie* N.F. 8, 1-10.

Rapp, A.; Annersten, L., 1969. Permafrost and Tundra Polygons in Northern Sweden. *Arctic Inst. North Amer. Montreal*, 65-91.

Rathjens, C., 1964. Ein Beitrag zur Frage der Solifluktionsgrenze in den Gebirgen Vorderasiens. *Zeitschrift für Geomorphologie* N.F. 9, 35-49.

Semmel, A., 1969. Verwitterungs- und Abtragungserscheinungen in rezenten Periglazialgebieten (Lappland und Spitzbergen). *Würzburger Geographische Arbeiten* 26, 1-82.

Shi Yafeng, Zheng Benxing, Li Shijie, 1992. Last glaciation and maximum glaciation in the Qinghai-Xizang (Tibet) Plateau: A controversy to M. Kuhle's ice sheet hypothesis.- *Z. Geomorph. N. F.,* Suppl.-Bd. 84, Berlin, Stuttgart, 19-35.

Shi Yafeng (Scientific advisor), Li Binyuan, Li Jijun (Chief editors), Cui Zhijiu, Zheng Benxing, Zhang Qingsong, Wang Fubao, Zhou Shangzhe, Shi Zuhui, Jiao Keqin, Kang Jiancheng (Editors). 1991. Quaternary Glacial Distribution Map of Qinghai-Xizang (Tibet) Plateau, Scale 1:3,000,000, *Science Press,* Beijing.

Shumskii, P. A., 1947. Energiia oledeneniia i zhizn lednikov (Energy of glacierization and the life of glaciers). Geografgiz, Moscow; transl. by W. Mandel, *The Stefansson Libary*, New York 1950, 1-60.

Shumskii, P. A., 1964. Principles of Structural Glaciology. translated from the Russian by D. Kraus; Dover Publ., Inc., New York, 1-497.

Svensson, H., 1967. Polygonal Ground and Solifluction Features, Photographic Interpretation and Field Studies in Northernmost Skandinavia. *Lund Stud. Geogr.*, A 40, 8-32.

Svensson, H., 1970. Pingos i yttre delen av Adventdalen. *Norsk Polarinst. Arbok 1969*, 168-174.

Troll, C., 1944. Strukturböden, Solifluktion und Frostklimate der Erde. *Geolog. Rdsch.* 34 (7/8), Stuttgart, 545-694.

Tricart, J. (ed.), 1967. Le modelé des régions périglaciaires. *Traité des Géomorphologie* Vol. 2, 1-512.

Troll, C., 1947. Die Formen der Solifluktion und die periglaziale Bodenabtragung. *Erdkunde* 1, 162-175.

Wang Shaoling, 1987. Frozen Ground and Periglacial Features in the Southeastern Part of Qinghai Province. In: Hövermann, J., Wenjing, W. (eds.): *Reports on the Northeastern part of the Quinhai-Xizang (Tibet)-Plateau by the Sino-West-German Scientific Expedition 1981*. Science Press, Beijing, China, 343-366.

Wahrhaftig, C., Cox, A., 1959. Rockglaciers in the Alaska Range.- *Bull. Geol. Soc. Amer.* 70, 383-436.

Washburn, A. L., 1956. Classification of patterned ground and review of suggested origins. *Bull. Geol. Soc. Amer.* 67 (7), 823-865.

Washburn, A. L., 1979. Geocryology. A survey of periglacial processes and environments, 1-406.

Zho Youwu and Guo Dongyin, 1982. Principal Characteristics of Permafrost in China. *J. Glaciology and Cryopedology*. 4 No. 1, 1-19; Lanzhou.

In: Permafrost
Editor: Oleg S. Pokrovsky

ISBN: 978-1-62948-830-1

Chapter 4

NUMERICAL MODELING OF HEAT AND WATER TRANSFER IN PERMAFROST-DOMINATED WATERSHEDS

Laurent Orgogozo*[*1], Oleg S. Pokrovsky[1], Yves Goddéris[1], Jérôme Viers[1], David Labat[1] and Anatoly S. Prokushkin[2]
[1]Geosciences Environment Toulouse, Toulouse, France
[2]V.N. Sukachev Institute of Forest, SB RAS, Akademgorodok, Russia

ABSTRACT

This chapter deals with the thermo-hydrological determinants of weathering fluxes in boreal watersheds, and with the insights on this matter that may be expected from a mechanistic modeling approach. Focus will be placed on the weathering fluxes in the permafrost dominated watersheds of the basaltic region of the Putorana Plateau in Central Siberia, one of the largest and most pristine boreal forested areas (Pokrovsky et al., 2005).

In the first part of this chapter a qualitative model of the thermo-hydrological functioning of the surfaces of the watersheds of this region will be presented. This qualitative model is based on quantitative geochemical measurements available in the literature (e.g. Pokrovsky et al., 2005, Prokushkin et al., 2007, Bagard et al., 2011). A discussion on

* Corresponding Author OMP-GET, 14, avenue Edouard Belin, 31400 Toulouse, France
Email: Laurent.Orgogozo@get.obs-mip.fr.

the couplings between water and energy fluxes and matter fluxes is briefly presented. The importance of phase changes and advective transfers is emphasised.

In the second part of this chapter, after a short review of published numerical tools dealing with the coupled transfer of water and energy in the active layer of permafrost areas, perspectives associated with the use of modern numerical techniques are discussed. A modeling efforts toward this direction is introduced, which has been initiated in the framework of OpenFOAM®, an open source tool box for computational fluid dynamics that allows multi-physic coupling and massively parallel computing (e.g. Weller et al., 1998). Finally, an example of a set of results is presented. These results illustrate the importance of the snow cover and the moss/soil organic horizon thickness on the thermo-hydrological behaviour of the soil active layer depth of this area.

Keywords: Permafrost, active layer, Central Siberia, weathering, mechanistical modeling, thermo-hydrological processes

INTRODUCTION

The weathering of continental silicated rocks is considered one of the main sinks of atmospheric carbon (Berner, 1992). Thus, in the context of global changes, the study of this phenomenon is a key point for our understanding of the dynamic of the green house effect. It has been shown that basaltic rock accounts for about 30% of the uptake of atmospheric carbon associated with the weathering of continental silicated rocks (Dessert et al., 2003). The Putorana Plateau is constituted of traps over an area of about 1 500 000 km^2 (Pokrovsky et al., 2005), so it is one of the largest areas in the world where basaltic rock weathering takes place. The observed global warming, at least partly due to anthropogenic carbon, seems to be particularly strong in boreal areas (e.g. IPCC, 1998). Such global warming is strongly suspected to induce the thawing of boreal permafrost, with a decrease in the permafrost extension and an increase in active layer thickness, even if these points are still a matter of debate (e.g. Serreze et al., 2000, Zhang et al., 2005, Frauenfeld et al., 2007). This could be the source of severe positive feedback on global warming through the release of organic matter contained in the frozen soil layers (Zimov et al., 2006, Khvorostyanov et al., 2008a). On the other hand, the impact of the potential increase in the thickness of the soil layers affected by weathering due to the prospective increase in the active layer thickness has not yet been assessed. The weathering process, as well as the dynamics of the active layer, in basaltic boreal areas such as the Putorana Plateau are therefore potentially important determinants of the global carbon cycle and the global warming scenario. Weathering and active layer dynamics are both closely linked to the thermo-hydrological regimes occurring in the soil of these regions, which may be subjected to major evolutions due to climate change. Indeed, the Putorana Plateau is an ideal place for studying weathering and permafrost dynamics in boreal basaltic areas, since it is one of the largest areas of this kind and since it is almost completely pristine (Pokrovsky et al., 2005).

As mentioned previously, the thermo-hydrological state of a soil exerts a strong control on its weathering and thus on the associated atmospheric carbon consumption (and also on other environmental aspects, such as water resources; e.g. Quinton et al. 2011). One can see, for example in Figure 1, a dependence of the weathering flux on the hydrological (run-off, computed) and thermal conditions (annual mean surface temperature, measured, 1982–1990) at 22 sites with silicated outcrops around the world. This illustration of the relationship between the thermo-hydrological state and weathering flux is taken from Beaulieu et al. (2010).

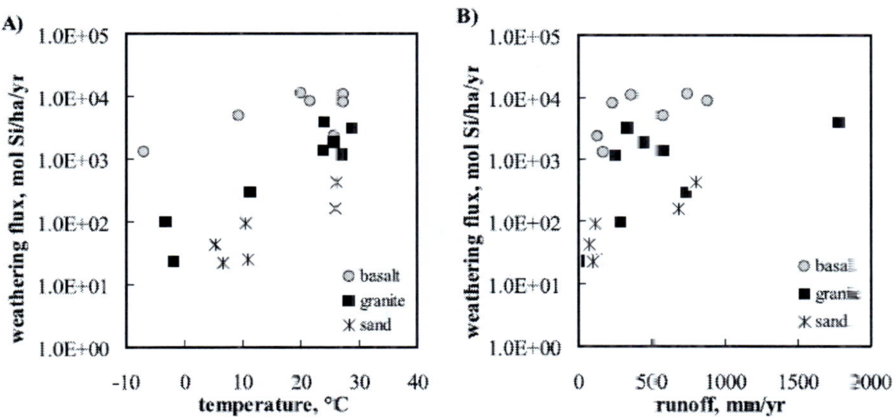

Figure 1. Illustration of correlations between weathering fluxes and (A) thermal regimes and (B) hydrological regimes for 22 representative silicated sites around the world (exerpt from Beaulieu et al., 2010).

On these two graphs one can see positive correlations between the weathering flux an (i) the surface temperature and (ii) the run-off. The first correlation may be associated with an increase in geochemical reaction kinetics with temperature. The second correlation may be related to an increase in the transfer of dissolved ions generated by the weathering of the soils. Thus both effects, thermal and hydrological, are important in weathering phenomenon. Moreover, there is a strong coupling between the hydrological and thermal processes in permafrost areas: snow melt and phase changes of the water in the active layer. The latent energy released by the freezing of water in the active layer at the beginning of winter and the latent energy needed for the thawing of this active layer in the hot season is a major coupling between thermal and hydrological processes. Indeed, it results in a strong dependency of the thermal dynamics of the active layer on its water content, which results in a damping of the thermal evolution of the soil (Farouki 1981, Hinkel et al., 2001).

This may have a strong impact even at continental scale (Takata and Kimoto, 2000, Poutou et al., 2004, Gouttevin et al., 2012). Other forms of coupling exist, for example through the insulation effect of the snowpack (Frauenfeld et al., 2004). Another example is the occurrence of advective heat transport in the active layer, which has been reported in surficial permafrost during spring flood, as shown in Figure 2 (Hinkel et al., 1997, 2001).

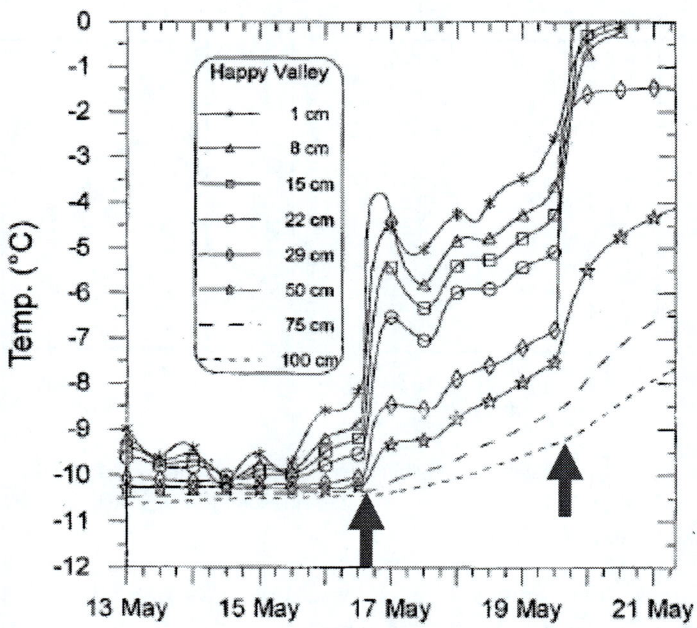

Figure 2. Monitoring of temperature profiles along depth during a spring flood period in a permafrost area of northern Alaska (extracted from Hinkel et al., 1997).

One can see on this graph some sharp jumps in temperature, even at 30 cm depth within the permafrost. These jumps take place at the same time as the primary snowmelt events of the year occur in this area. These fast jumps in temperature at tens of centimetres from the surface cannot be generated by thermal diffusion, they can only be explained by advective transfers of heat through the soil layers. Thus a modeling of the active layer thickness of the permafrost and of the potential impact of global warming on its dynamic needs to fully couple thermal and hydrological effects in order to represent the physics at stake. Indeed, current continental scale modeling of the carbon cycle in boreal areas does take into account such a coupling with conceptual approaches of modeling, but they still have high parameter sensitivities (e.g. Zhuang et al., 2003, Khvorostyanov et al., 2008b). Performing modeling studies of element cycles with a fully mechanistic modeling approach at smaller scales would allow stronger constraints on the processes involved to be obtained. Indeed, the need for developing such mechanistic modeling tools is now recognised (e.g. Frampton et al., 2011, Grenier et al., 2012).

Putorana Plateau watersheds have been studied for years in order to constrain weathering phenomenon in basaltic boreal areas (e.g. Pokrovsky et al., 2005, Prokushkin et al., 2007, Bagard et al., 2011). It has been an opportunity to build a qualitative model of the thermo-hydrological processes at work on the basis of quantitative geochemical field measurements. Thus it would be the ideal place to develop fully mechanistic modeling studies of the thermo-hydrological dynamics and of the weathering process at a small watershed scale, because it is a very pristine area and because of the careful characterisation and monitoring of this area. In the first part of this chapter we will focus on the current qualitative knowledge on thermo-hydrological regimes encountered in this area. In the second part, we will deal with the mechanistic modeling approaches of the thermo-hydrological processes in permafrost dominated areas, with a brief overview of the literature, a discussion on the use of modern numerical techniques in this context, and an illustrative example of results.

THERMO-HYDROLOGICAL PROCESSES IN THE NEAR-SURFACE PERMAFROST OF THE PUTORANA PLATEAU

This summary of current knowledge about the thermo-hydrological regime of the watersheds in the Putorana Plateau is based on several studies conducted on catchments within the Nizhniya Tunguska River and in the Kochechum River watersheds. Temporally, three main phases may be determined in the thermo-hydro-geochemical functioning of Putorana Plateau watersheds: winter season (roughly from October to the beginning of May), spring flood (mid-May and June), and summer-fall season (July to September) (Prokushkin et al., 2009). The Central Siberian Plateau is mainly covered with taïga (below 800 m a.s.l.), forest-tundra (between 800 and 1000 m a.s.l.) and tundra (above 1000 m a.s.l.), and it is a permafrost dominated area (from discontinuous permafrost in the south to thick continuous permafrost in the north) (Brown et al., 1998). The hydrological compartments that may be considered are the following: surficial organic layer (including both moss-lichen stratum and soil organic horizons), soil active layer (the layer thawed during the frost-free period), surface water (rivers), and deep groundwater (i.e., brines underneath the permafrost) (Pokrovsky et al., 2005).

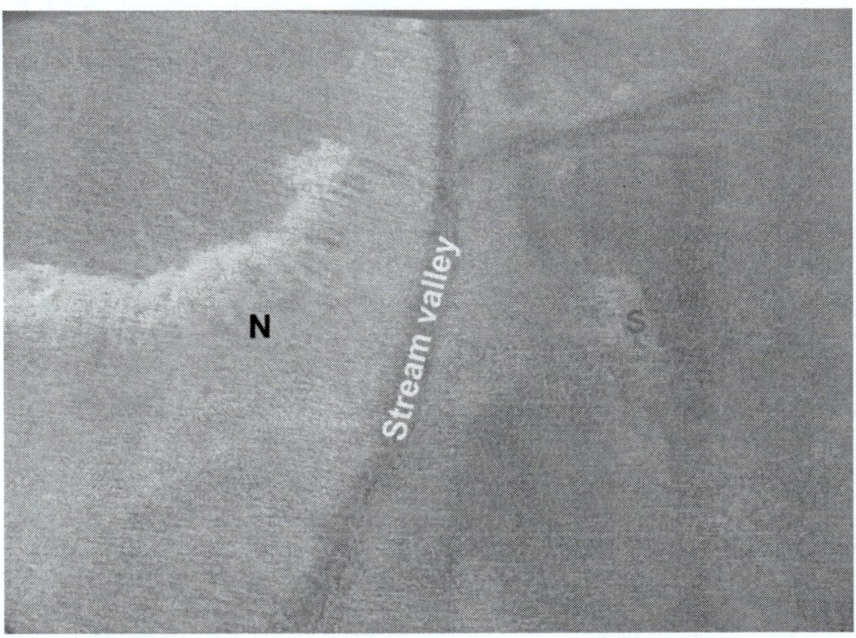

Figure 3. North and south aspected slopes in a valley of the Putorana Plateau.

The study of the time variations and spatial variations of the major cations dissolved in the riverwaters associated with these different compartments shows that the most important flux of matter associated with weathering occurs during spring flood, although weathering could occur prior to the spring flood (e.g. under freeze-thaw cycles, in this case those products are released to rivers by snow-melt waters).

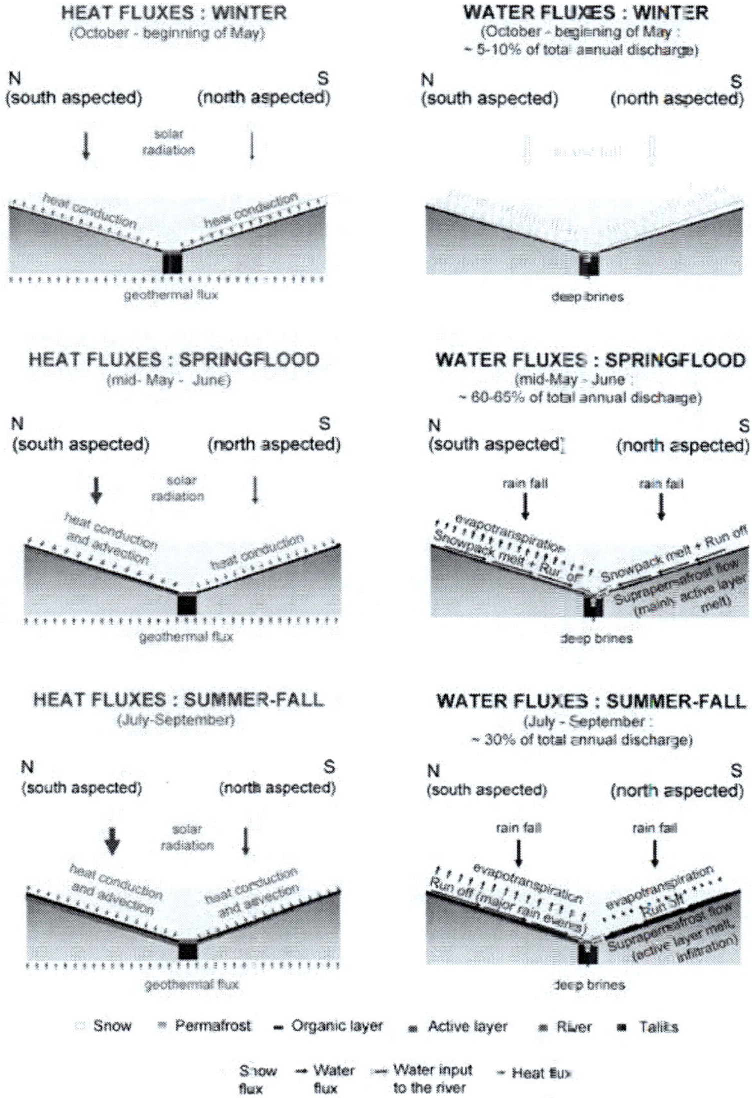

Figure 4. Scheme of thermo-hydrological functionning in Putorana Plateau: couplings between thermal and hydrological fluxes and north aspected / south aspected slope contrasts (mainly according to Pokrovsky et al., 2005, Prokushkin et al., 2007, Bagard et al., 2011).

For example, more than 50% of the total export of dissolved cations in the river occurs during spring flood (Pokrovsky et al., 2005). Trace elements studies also indicate an intense leaching of the organic layer during spring flood (Pokrovsky et al., 2006). Quantitative and statistic analysis of dissolved organic matter in the riverwater indicate a clear difference in thermo-hydrological behaviour between the north aspected slopes and the south aspected slopes (Prokushkin et al., 2007), a difference which may be driven for example by the difference in solar radiation input on both slopes. This difference in thermo-hydrological conditions between north and south aspected slopes is even visible in the landscapes of this region, as shown in Figure 3.

The occurrence of deep brine supplied to the river through taliks has been shown to be a determinant for the chemical composition of the riverwater in winter (Bagard et al. 2011). Overall, the thermo-hydrological functioning of the watershed of the Putorana Plateau can be sketched from these studies as shown in Figure 4.

The information contained in this scheme may be summarised as follows:

- Winter time: the soil water is completely frozen, and the meteoric waters are in the form of snow. One can observe a thicker snow pack in the north aspected slope, due to smaller snow sublimation than in the south aspected slope, in relation to the difference in solar radiation heat supplies. The water flux in rivers is at its lowest level of the year (5–10% of the total annual discharge, according to Pokrovsky et al., 2005). The water flowing in the river is almost exclusively of deep origin, with a below-permafrost brine supply through taliks.
- Spring flood: melting of the snow pack and beginning of the thawing of the active layer, mainly in the south aspected slopes, even if first surficial ice-rich permafrost horizons of the north aspected slopes start to melt at the end of the spring flood. The meteoric waters are liquid, and then migrate to the river through organic layer leaching on both slopes. The infiltrations of rain waters or snowmelt waters may be the cause of heat advection to the frozen ground through the active layer, and thus increase the thawing rate. This advective heat transfer first occur in the south aspected slopes in which such an active layer exists earlier, and then at the end of spring flood in the north aspected slopes. Transpiration starts with vegetal activity, which is more intensive in south aspected slopes at this time of the year. Then, even if the thawing of permafrost is stronger in the south aspected slopes,

most of these waters are transpirated and few of them reach the stream. The largest part of the total annual discharge of the river occurs (60–65%, according to Pokrovsky et al., 2005), and the water in the river is a mix between run-off waters and snow melting from both slopes, and, at the end of the spring flood, soil ice melting of the north aspected slope (and a small proportion of deep brines).

– Summer: During the summer season, an active layer develops on both slopes, and it is deeper in the south aspected slope. During this period, the water in the river is a mixture of run-off water and north aspected slope infiltration water and active layer ice melting (and a small proportion of deep brines). In the south aspected slopes infiltration occurs, but precipitation is not sufficient to reach the unpermeable layers and then to be released to rivers by lateral flux. Due to a thicker active layer and stronger canopy effects than in the north aspected slopes, almost all the infiltrated waters are re-emitted to the atmosphere through evapotranspiration. Even the run-off water input to the river from these south aspected slopes is limited to major rain events, because of the higher water storage capacity of the deeper thawed soil than in the north aspected slopes.

Several remarks may be made at this point. First, the strong coupling between thermal and hydrological processes is obvious: the thickness of the snow pack in relation to solar exposure, advective transport of heat in the active layer, phasing of the origin of soil waters in rivers in relation to the active layer thickness differences … Second, one can see that we have highly-contrasted thermo-hydrological functioning, both temporally ("flashy" transfers during spring flood) and spatially (north aspected slopes versus south aspected slopes). Finally, the export of water (and thus the associated flux of organic matter) from the south aspected slopes is smaller during the summer season than in spring flood, and than that observed in the north aspected slopes. If one consider that the "south aspected slope's behaviour" may be regarded as the possible future behaviour of the north aspected slopes in the context of climate warming, then it could be anticipated that in the future these large areas will uptake more atmospheric carbon than now. It would be an opposite effect to the one of a potential release to rivers of organic matter currently entrapped in the permafrost surficial layers. Then the impact of various scenarii of climate change (dry or wet, more or less strong, etc.) need to be carefully assessed due to the complexity of the considered system, with strong couplings, contrasted behaviours, and contradictory effects.

Mechanistical Modeling of Thermo-Hydrological Behaviour of Soils in Permafrost-Dominated Areas

Quantitative modeling of the thermo-hydrological processes may allow a better understanding of the active layer dynamics, and its possible evolution in the context of climate change. Such complex systems (coupled, contrasted, evolving) require an a priori approach to modeling (as far as possible), in order that the basic assumptions of the modeling do not hide the true functioning of the studied system. To the knowledge of the authors, the most a priori method of modeling is the mechanistical modeling approach, i.e., the resolution of the equations of conservation obtained from the continuum mechanics. It is the modeling approach with the most limited and precisely identified a priori assumptions on the functioning of the modelled system. This leads to predictive modeling under the assumptions of the model. It allows for example to test the likelihood from a physical point of view of a scheme of functionning of the studied system through numerical experiments. An example of such an approach is shown below. The predictivity of mechanistic modeling is a major strength of these approaches compared to conceptual modeling (Li et al., 2008, Yeh et al., 2011, Miller et al., 2013). Conceptual modeling approaches are based on the calibration of transfer functions that empirically relate the input of the considered system (for example rain and snow falls and air temperature) and the desired output (active layer thickness or water content, for instance). In the case of conceptual models, the study of a system in evolving conditions is very difficult, since there is no theoritical reason that a calibration made with given input data will still be pertinent with other input data (for example in the case of climate change) (van der Ploeg et al., 2012). However the drawbacks of the mechanistical approach are that it requires a good knowledge of the physical and geometrical characteristics of the system under consideration (Carrera-Hernandez et al., 2012), and that it leads in general to systems of equations that are not analytically solvable. Thus numerical approximate resolution is required, with potentially high costs in terms of computation power and computation time (Miller et al., 2013).

In the field of a freezing soil-water system, numerous works have been done to establish one-dimensional mechanistic models (e.g. Kennedy and Lielmezs, 1973, Harlan, 1973, Guymon and Luthin, 1974, Jame and Norum, 1980, Mu and Ladanyi, 1987, Seregina, 1989, Hansson et al., 2004). They consist in establishing governing equations for water and heat fluxes within a variably saturated and variably frozen soil column, as well as the numerical procedure to solve these equation. These one-dimensional models are built on the assumption that the humid air phase is at a constant pressure (atmospheric pressure) within the soil, leading to a one equation model of mass (water) transfer, namely Richards' equation. In this equation, gravity and capillarity are taken into account. The equation of heat transfer takes into account the conductive and advective transport of heat. These two equations are coupled through water phase changes and the advective transport of heat, as well as through the dependency of the hydrodynamic and thermal properties of the soil on its liquid and ice water contents. Most of these models neglect the coupling between soil mechanics and thermo-hydrological fluxes, such as ice lenses formation, frost heave, etc. (e.g. Kennedy and Lielmezs, 1973, Harlan, 1973, Guymon and Luthin, 1974, Jame and Norum, 1980, Seregina, 1989, Hansson et al., 2004). Nevertheless, some works do take into account these soil mechanics – thermo-hydrological fluxes couplings (e.g. Mu and Ladanyi, 1987). The study of these additional couplings is beyond the scope of this chapter, but one can find an introduction to these problems in Andersland and Ladanyi (1994).

One-dimensional modeling does not consider topographical effects, at least directly, and only allows a schematic (layered) description of the pedology and geology of the studied cases. Thus efforts have been made to build two-dimensional models. For example, some works have developed two-dimensional simulations in the context of both martian and terrestrial permafrost, using a two-phase flow approach in order to deal with peculiar conditions of pressure and temperature encountered on the surface of Mars (Painter, 2011). This tool has been successfully used to deal with the two-dimensional modeling of terrestrial permafrost (Frampton et al., 2011, Sjöberg et al., 2012.). Two-dimensional tools for saturated freezing soils (Grenier et al., 2012, Jost, 2012) have also been developed to study the behaviour of the Paris Basin at the time scale of the ice ages. Nevertheless, no fully three-dimensional tools have yet been developed. One of the major obstacles to conducting three-dimensional thermo-hydrological modeling of variably saturated and variably frozen soils is the huge computation power and computation time that it would require. Indeed, the numerical resolution of such sets of equations requires small time steps and fine spatial resolution (e.g. Romanovsky et al., 1997). An open source tool developed in the area of computational fluid dynamics, namely OpenFOAM® (Jasak, 1996, Weller et al., 1998, www.openfoam.com), has recently shown its capability in dealing with modeling with intensive numerical costs in the field of soil hydrodynamics (Orgogozo et al., 2012, 2013, Submitted), mainly through the use of massively parallel computation. A first attempt to apply this tool to the study of freezing soils is currently being undertaken, and a little one-dimensional example of results is shown below.

With this example we intend to illustrate the importance of the insulation effect of the snow cover and organic layer on the soil active layer depth dynamics in central Siberia permafrost dominated areas, with a very simplified physical approach. Our approach is the following: if we neglect the snow pack effects and the presence of an organic layer, do we find realistic (i.e., not too far from observations) thermo-hydrological behaviour in our soil column. So we consider a loam column of 4 m in thickness, partially saturated with water, with no organic layer. We have to seek a sinusoidal air temperature annual variation that leads to an active period and a maximum active layer thickness that are comparable to those observed on Putorana Plateau watersheds (June to September and about 1 m, respectively), without considering the effect of the snow pack (i.e., by considering that the temperature at the top of the soil is equal to air temperature). The temperature at the bottom of the soil column has been assumed to be equal to the annual mean of the air temperature (-5°C). For this illustrative example we have not considered any water flow, so that no advective transfer of heat occurs. The resulting simplified equation system is shown below.

Heat transfer equation:

$$\nabla\left(\mathbf{K}_{\mathbf{T}}^{*}.\nabla T\right)=C_{T}^{*}\frac{\partial T}{\partial t}-L_{T}^{*}\frac{\partial \theta_{i}}{\partial t} \tag{1}$$

Phase change equation:

$$\theta_{i}=\begin{cases}0 \text{ if } T>T_{f}+\Delta T\text{, i.e. } \theta_{tot}=\theta_{i}+\theta_{l}=\theta_{l} \text{ with } \theta_{i}=0\\ \dfrac{\theta_{l}}{\dfrac{1}{2}+\dfrac{\left(T-T_{f}\right)}{2\Delta T}} \text{ if } T_{f}+\Delta T\geq T\geq T_{f}-\Delta T\text{, i.e. } \theta_{tot}=\theta_{i}+\theta_{l}\\ \theta_{tot} \text{ if } T_{f}-\Delta T>T\text{, i.e. } \theta_{tot}=\theta_{i}+\theta_{l}=\theta_{i} \text{ with } \theta_{l}=0\end{cases} \tag{2}$$

Effective thermal conductivity:

$$\mathbf{K}_{\mathbf{T}}^{*} = (1-\phi)\mathbf{K}_{\mathbf{T},soil} + \theta_l \mathbf{K}_{\mathbf{T},water} + \theta_g \mathbf{K}_{\mathbf{T},ice} \quad (3)$$

Effective volumetric calorific capacity:

$$C_T^{*} = (1-\phi)C_{soil} + \theta_l C_{water} + \theta_g C_{ice} \quad (4)$$

Effective volumetric latent heat of fusion of soil water:

$$L_T^{*} = L_{fusion,ice}\,\theta_l / (1-\phi) \quad (5)$$

Here, stands for the soil temperature [K], is the temperature of the fusion of ice [K], is the icy water volumetric content in the soil [$m^3.m^{-3}$], is the liquid water volumetric content in the soil [$m^3.m^{-3}$], is the total water volumetric content in the soil [$m^3.m^{-3}$], is the effective thermal conductivity of the soil [$J.s^{-1}.K^{-1}.m^{-1}$] (i.e., the thermal conductivity of the porous medium constituted by soild grains, air, the liquid water and icy water), is the effective volumetric calorific capacity of the soil [$J.m^{-3}.K^{-1}$], is the effective volumetric latent heat of fusion of soil ice [$J.m^{-3}$], is the thermal conductivity of the component i of the soil [$J.s^{-1}.K^{-1}.m^{-1}$], is the volumetric calorific capacity of the component i of the soil [$J.m^{-3}.K^{-1}$] (i may stand for solid, water or ice), is the porosity of the soil [$m^3.m^{-3}$], and is the volumetric latent heat of free ice [$J.m^{-3}$]. Several remarks may be made here. First, one can see that the evaluation of the effective thermal conductivity and the effective calorific capacity of the soil are based on a local thermal equilibrium assumption, following for instance Guymon and Luthin (1974). Second, the evaluation of the effective latent heat of fusion of the soil ice, which is influenced by interfacial processes, interaction with the solid skeleton, etc., is done with the simple empirical formulae of Kersten (1949) – Eq. (5). Finally, one can note that we use a linear interpolation for the ice content around the temperature of the fusion of ice T_f, with a small (i.e., $T_f >>$) parameter [K] for the interval of linearisation. Figure 5 presents the table of the classical order of magnitude considered for the various physical parameters of the model.

	Soil	Solid grains	Liquid water	Ice
Thermal conductivity [MJ.s^{-1}.K^{-1}.m^{-1}]		7.5*10^{-7}	6*10^{-7}	2.1*10^{-6}
Volumetric calorific capacity [MJ.m^{-3}.K^{-1}]		1.53	4.18	1.89
Latent heat of fusion [MJ.m^{-3}]				305.36
Porosity [-]	0.43			
Total water content [-]	0.33			

Figure 5. Typical values of model parameters discussed in this study.

Then, Figure 6 shows a summary of the obtained results, with the considered climatic input signal and the evolutions of the active layer, water state and temperature within the soil column over a year.

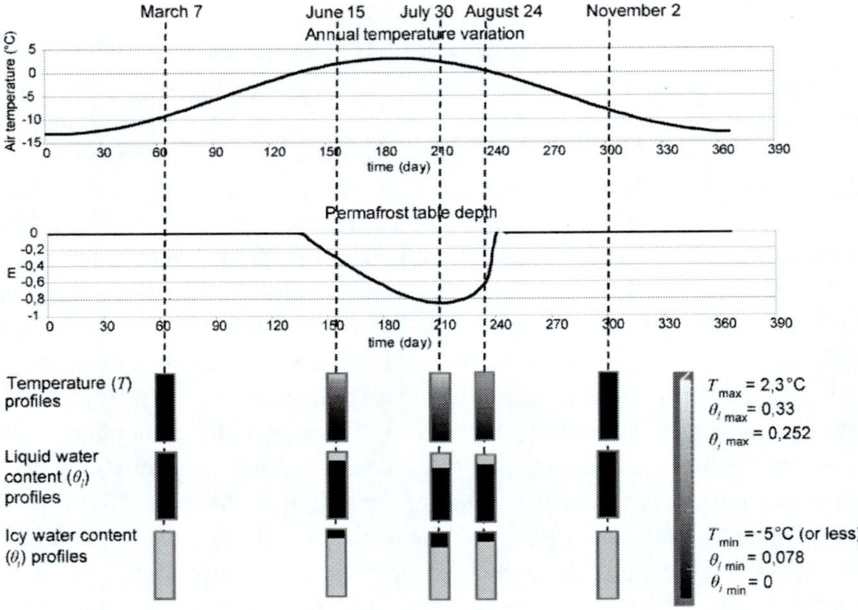

Figure 6. Permafrost table depth, and some temperature, liquid water content and icy water content profiles within a 4 m thick loess soil column submitted to a sinusoidal annual variation in temperature at its surface (mean : -5 °C, amplitude: 16 °C). Values obtain by numerical resolution of equations (1) to (5) using a homemade OpenFOAM® solver.

One can see that we obtain a maximum active layer thickness of slightly less than one metre with a surface temperature signal far smoother than the one for the region of the Putorana Plateau with a mean annual surface temperature at Tura station in the Putorana Plateau: -9°C, and rough amplitude

: 53 °C. Thus we can see that neglecting the organic layer and the snow pack insulation effect leads to strong discrepancies between the theoretical behaviour obtained by modeling and that observed. It is numerical evidence of the importance of the organic layer and of the snow pack in the thermo-hydrological behaviour of the Putorana Plateau catchment. Also note the temporal asymmetry of the active layer's thickness variation, due to the thermal inertia of the soil column. It shows that the phenomena are more complicated than a simple linear dependency of active layer thickness on surface temperature, and thus justifies the need to use mechanistic approaches. The computation shown here is one-dimensional, but the solvers developed in the framework of OpenFOAM® are by default 3D solvers, which can be applied to 1D or 2D modeling by using peculiar empty boundary conditions..

Conclusion

As shown by the study of weathering fluxes in the Putorana Plateau, the thermo-hydrological behaviour of a boreal catchment in a permafrost dominated area is a key parameter in the export of matter from the catchment. It is then an important stake for the study of the impact of climate change on the export of matter from boreal areas and on the associated potential feedback. The complexity of the physics involved here (coupling between heat and water fluxes, highly contrasted regimes both seasonally and spatially – north aspected slope/south aspected slope, antagonistic effects of climate change expected ...) requires a quantitative and predictive modeling approach, if one wants to infer the reaction of these systems to global changes. Mechanistic modeling is recognised as the "golden standard" in the study of hydrosystems dynamics (Miller et al., 2013), and many efforts have been made in the past to build efficient tools to study the physical behaviour of freezing soil-water systems. Nevertheless these works were limited in applicability due to a lack of computation power. The generalisation of the use of massively parallel computing may allow the community of modellers of thermo-hydrological behaviour of permafrost to overcome this difficulty, and to move towards the predictive modeling of the response of these complex systems to climate changes.

Acknowledgments

The financial support from a BIO-GEO-CLIM Mega-grant from the Ministry of Education and Science of the Russian Federation and Tomsk State University (No 14.B25.31.0001), ANR "Arctic Metals", and GDRI CAR WET SIB, are acknowledged.

References

Andersland O.B., Ladanyi B., 1994. An introduction to frozen ground engineering. (New York: Chapman & Hall).

Bagard M.-L., Chabaux F., Pokrovsky O.S., Viers J., Prokushkin A.S., Stille P., Rihs S., Schmitt A.-D., Dupré B., 2011. Seasonal variability of element fluxes in two Central Siberian rivers draining high latitude permafrost dominated areas. *Geochimica et Cosmochimica Acta* 75, 3335-3357.

Beaulieu E., Goddéris Y., Labat D., 2010. Impact of atmospheric CO2 levels on continental silicate weathering. *Geochemistry Geophysics Geosystems* 11(7), doi: 10.1029/2010GC003078.

Berner R.A., 1992. Weathering, plants, and the long-term carbon cycle. *Geochimica et Cosmochimica Acta* 56(8), 3225-3231.

Brown J., Ferrians O.J. Jr, Heginbottom J.A., Melnikov E.S., 1998 (Revised February 2001). Circum-Arctic Map of Permafrost and Ground Ice Conditions (Boulder, CO: National Snow and Ice Data Center/World Data Center for Glaciology).

Carrera-Hernández J.J., Smerdon B.D., Mendoza C.A., 2012. Estimating groundwater recharge through unsaturated flow modeling: sensitivity to boundary conditions and vertical discretization. *Journal of Hydrology* 452-453, 90-101.

Dessert C., Dupré B., Gaillardet J., Françoid L.M., Allègre C.J., 2003. Basalt weathering laws and the impact of basalt weathering on the global carbon cycle. *Chemical Geology* 2002(3-4), 257-273.

Farouki O.T., 1981. The thermal properties of soils in cold regions. *Cold Regions Science and Technology* 5, 67-75.

Frampton A., Painter S., Lyon S.W., Destouni G., 2011. Non-isothermal, three-phase simulations of near-surface flows in a model permafrost system under seasonal variability and climate change. *Journal of Hydrology* 403(3-4), 352-359.

Frauenfeld O.W., Zhang T., Barry R.G., Gilichinsky D., 2004. Interdecadal changes in seasonal freeze and thaw depths in Russia. *Journal of Geophysical Research* 109, doi: 10.1029/2003JD004245

Frauenfeld O.W., Zhang T., McCreight J.L., 2007. Northern hemisphere freezing/thawing index variations over the twentieth century. *International Journal of Climatology* 27, 47-63.

Gouttevin I., Krinner G., Ciais P., Polcher J., Legout C., 2012. Multi-scale validation of a new soil freezing scheme for a land-surface model with physically-based hydrology. *The Cryosphere* 6, 407-430.

Grenier C., Régnier D., Mouche E., Benabderrahmane H., Costard F., Davy P., 2012. Impact of permafrost development on groundwater flow patterns: a numerical study considering freezing cycles on a two-dimensional vertical cut through a generic river-plain system. *Hydrogeology Journal*, doi: 10.1007/s10040-012-0909-4.

Guymon G.L., Luthin J.N., 1974. A coupled heat and moisture transport model for arctic soils. *Water Resources Research* 10(5), 995-1001.

Hansson K., Šimůnek J., Mizoguchi M., Lundin L.-C., van Genuchten M.Th., 2004. Water flow and heat transport in frozen soil: numerical solution and freeze-thaw applications. *Vadose Zone Journal* 3, 693-704.

Harlan R.L., 1973. Analysis of coupled heat-fluid transport in partially frozen soil. *Water Resources Research* 9(5), 1314-1323.

Hinkel K.M., Outcalt S.I., Taylor A.E., 1997. Seasonal patterns of coupled flow in the active layer at three sites in northwest North America. *Canadian Journal of Earth Sciences* 34, 667-678.

Hinkel K.M., Paetzold F., Nelson F.E., Bockheim J.G., 2001. Patterns of soil temperature and moisture in the active layer and upper permafrost at Barrow, Alaska: 1993-1999. *Global and Planetary Change* 29, 293-309.

IPCC, 1998. The Regional Impacts of Climate Change: An Assessment of Vulnerability (Cambridge: Cambridge University Press).

Jame Y.-W., Norum D.I., 1980. Heat and mass transfer in a freezing unsaturated porous medium. *Water Resources Research* 16(4), 811-819.

Jasak H, 1996. Error Analysis and Estimation for the Finite Volume Method with Applications to Fluid Flows. PhD Thesis, Imperial College of Science, Technology and Medicine, London.

Jost A., 2012. 2D modeling of fluid flow and heat transfer in the Paris basin, France: an assessment of the impact of permafrost during the last glacial/interglacial cycle. Abstracts of Quaternary International 227.

Kenedy G.F., Lielmezs J., 1973. Heat and mass transfer of freezing water–soil system. *Water Resources Research* 9(2), 395-400.

Kersten M.S., 1949. Laboratory research for determination of the thermal properties of soils, Technical Report 23. (Minneapolis: Engineering Experiment Station, University of Minnesota).

Khvorostyanov D.V., Ciais P., Krinner G., Zimov S.A., 2008a. Vulnerability of East Siberia's frozen carbon stores to future warming. *Geophysical Research Letters* 35, doi: 10.1029/2008GL033639.

Khvorostyanov D.V., Ciais P., Krinner G., Zimov S.A., Corradi Ch., Guggenberger G., 2008b. Vulnerability of permafrost carbon to global warming. Part II: sensitivity of permafrost carbon stock to global warming. *Tellus* 60B, 265-275.

Li Q., Unger A.J.A., Sudicky E.A., Kassenaar D., Wexler E.J., Shikaze S., 2008. Simulating the multi-seasonnal response of a large-scale watershed with a 3D physically-based hydrologic model. *Journal of Hydrology* 357, 317.

Miller C.T., Dawson C.N., Farthing M.W., Hou T.Y., Huang J., Kees C.E., Kelley C.T., Langtangen H.P., 2013. Numerical simulation of water resources problems: models, methods, and trends. *Advances in Water Resources* 51, 405-437.

Mu S., Ladanyi B., 1987. Modeling of coupled heat, moisture and stress field in freezing soil. *Cold Regions Science and Technology* 14, 237-246.

Orgogozo L., Renon N., Hénon F., Soulaine C., 2012. Application of OpenFOAM to transfers in soils: first developments and perspectives. 7th OpenFOAM® Workshop, Darmstadt, Germany.

Orgogozo L., Labat D., Renon R., Tomer S.K., Soulaine C., Hénon F., Ababou R., Quintard M., 2013. A massively-parallel OpenFOAM® solver for Richards equation: towards mechanistic modeling of transfers at the watershed scale. 5th InterPore Conference, Prague, Czech Republic.

Orgogozo L., Renon N., Soulaine C., Hénon F., Tomer S.K., Labat D., Pokrovsky O.S., Sekhar M., Ababou R., Quintard M., Submitted. An open source parallel solver for Richards equation: mechanistic modeling of water fluxes at the watershed scale.

Painter S.L., 2011. Three-phase numerical model of water migration in partially frozen geological media: model formulation, validation, and applications. *Computers & Geosciences* 15, 69-85.

Pokrovsky O.S., Schott J., Kudryavtzev D.I., Dupré B., 2005. Basalt weathering in Central Siberia under permafrost conditions. *Geochimica et Cosmochimica Acta* 69(24), 5659-5680.

Pokrovsky O.S., Schott J., Dupré B., 2006. Trace element fractionation and transport in boreal rivers and soil porewaters of permafrost-dominated basaltic terrain in Central Siberia. *Geochimica et Cosmochimica Acta* 70, 3239-3260.

Poutou E., Krinner G., Genthon C., de Noblet-Ducoudré N., 2004. Role of soil freezing in future boreal climate change. *Climate Dynamics* 23, 621-639.

Prokushkin A.S., Gleixner G., McDowell W.H., Ruehlow S., Schulze E.-D., 2007. Source- and substrate-specific export of dissolved organic matter from permafrost-dominated forested watershed in central Siberia. Global *Biogeochemical Cycles* 21, doi:10.1029/2007GB002938.

Prokushkin A.S., Kawahigashi M., Tokareva I.V., 2009. Global warming and dissolved organic carbon release from permafrost soils. In: Permafrost Soils, Soil Biology 16, R. Margesin (ed.) (Berlin/Heidelberg: Springer-Verlag), doi: 10.1007/978-3-540-69371-0.

Quinton W.L., Hayashi M., Chasmer L.E., 2011. Permafrost-thaw induced land-cover change in the Canadian subarctic: implications for water resources. *Hydrological Processes* 25, 152-158.

Romanovsky V.E., Osterkamp T.E., Duxbury N.S., 1997. An evaluation of three numerical models used in simulations of the active layer and permafrost temperature regimes. *Cold Regions Science and Technology* 26, 195-203.

Seregina N.V., 1989. Some of the mathematical models used in geocryology and methods of their numerical solution (in Russian). *Geocryological Investigations* (Moscow: Moscow State University Press) pp. 243-246.

Serreze M.C., Walsh J.E., Chapin III F.S., Osterkamp T., Dyurgerov M., Romanovsky V., Oechel W.C., Morison J., Zhang T., Barry R.G., 2000. Observational evidence of recent change in the northern high-latitude environment. *Climatic Change* 46, 159-207.

Sjöberg Y., Frampton A., Lyon S.W., 2012. Using streamflow characteristics to explore permafrost thawing in northern Swedish catchments. *Hydrogeology Journal,* doi: 10.1007/s10040-012-0932-5.

Takata K., Kimoto M., 2000. A numerical study on the impact of soil freezing on the continental-scale seasonal cycle. *Journal of the Meteorological Society of Japan* 78(3), 199-221.

Van der Ploeg M.J., Haldorsen S., Leijnse A., Heim M., 2012. Subpermafrost groundwater systems: dealing with virtual reality while having virtually no data. *Journal of Hydrology*, http://dx.doi.org/10.1016/j.jhydrol.2012.08.046

Weller H.G., Tabor G., Jasak H., Fureby C., 1998. A tensorial approach to computational continuum mechanics using object-oriented techniques. *Computers in Physics* 12(6), 620-631.

Yeh G.-T., Shih D.-S., Cheng. J.-R.C., 2011. An integrated media, integrated processes watershed model. *Computers & Fluids* 45, 2.

Zhang T., Frauenfeld O.W., Serrez M.C., Etringer A., Oelke C., McCreight J., Barry R.G., Gilinchinsky D., Yang D., Ye H., Ling F., Chudinova S., 2005. Spatial and temporal variability in active layer thickness over the Russian Arctic drainage basin. *Journal of Geophysical Research* 110, doi: 10.1029/2004JD005642.

Zhuang Q., McGuire A.D., Melillo J.M., Clein J.S., Dargaville R.J., Kicklighter D.W., Myneni R.B., Dong J., Romanovsky V.E., Harden J., Hobbie J.E., 2003. Carbon cycling in extratropical terrestrial ecosystems of the Northern Hemisphere during the 20th century: a modeling analysis of the influences of soil thermal dynamics. *Tellus* 55B, 751-776.

Zimov S.A., Schuur E.A.G., Chapin III F.S., 2006. Permafrost and the global carbon budget. *Science* 312, 1612-1613.

In: Permafrost
Editor: Oleg S. Pokrovsky
ISBN: 978-1-62948-830-1

Chapter 5

REMOTE STUDY OF THERMOKARST LAKES DYNAMICS IN WEST SIBERIAN PERMAFROST

Yury Polishchuk[1,2], **Sergey Kirpotin**[1] **and Natalia Bryksina**[3,4]*
[1]Tomsk State University, Tomsk, Russia
[2]Institute of Petroleum Chemistry, Tomsk, Russia
[3]Kant Baltic State University, Kaliningrad, Russia
[4]Ugra State University, Khanty-Mansiysk, Russia

ABSTRACT

The features of the development of thermokarst processes on the territory of the frozen bog-lake systems in today's global warming are discussed. Thermokarst lakes are considered to be a suitable indicator of geocryological changes in permafrost landscapes using remote sensing. Methodological issues of remote measurement of lakes' areas based on optical and radar satellite images are discussed. The intra-seasonal dynamics of thermokarst lakes' areas was studied using radar images from ERS-2 and ENVISAT. Studies of long-term dynamics of thermokarst based on space image from Landsat were carried out at test sites located in different permafrost types and in different landscape zones.

Keywords: Permafrost, thermokarst lakes, remote sensing, satellite images, global warming

INTRODUCTION

Global climate change is a generally acknowledged fact, and expected consequences of climatic changes may affect many branches of the economy and social sphere in most countries in the world. The given problem is directly connected with the problem of energy security, the maintenance of which substantially depends on the sustainable operation and development of the oil and gas industry of Russia and other countries. So for the last three decades study of changes in permafrost state has become a very important problem, the solving of which is impossible without using remote sensing data because of hard access to the northern territories of West Siberia.

Peatlands situated in a permafrost zone and consisting of thin layers of frozen peat are a very sensitive indicator of climate changes, as any changes of climate in the direction of warming lead to increased activity in a thermokarst process over extensive areas (Smith et al., 2005; Samsonov et al., 2007). According to the investigations carried out by many scientists, for instance Kirpotin et al. (2009), the degree of thermokarst activity has unusually increased, and for the last decades has become extremely high. Thermokarst lakes which originate from ground subsidence as a result of fossil ice and frozen rocks melting were chosen as the most convenient geomorphological indicators of the global warming influence on the state of permafrost rocks for remote study of thermokarst dynamics.

Intensification of geocryological processes in the permafrost caused by climate changes will be accompanied by enhanced impact on infrastructure facilities and other buildings in arctic regions resulting in the increase in geoecological risks and economic damage. In this case getting predictive assessments of changes in the permafrost state under the influence of climate changes is an important task that can be performed by means of using experimental data concerning permafrost state changing. Collecting such data because of hard access to West Siberian wetland is impossible without remote sensing ground surface.

Remote studying of thermokarst lakes' state changes is carried out in Scandinavia, Siberia, Alaska and Canada. Most studies focus on qualitative analysis of the situational changes. Thus, Kravtsova and Bystrova's paper (2009) gives the results of the research in lake area changes in 9 test sites in the permafrost in Russia. The assessment of lake area changes was based on comparing the changes of coastal frontiers of the lakes in optical non-simultaneous satellite images. The published works devoted to the analysis of quantitative data concerning lakes' areas collected by means of remote techniques are rare, and they do not study the dynamics of thermokarst lake fields in the permafrost. The results of comparison of Landsat satellite images made in 1973 and 1998 (Smith et al., 2005) showed the changes of the total area of lakes in the permafrost in West Siberia. Along with this it was found that out that in a continuous permafrost lakes' areas mainly increase while in more southerly zones of discontinuous and sporadic permafrost they tend to decrease.

As one can infer from the aforementioned the research focused on thermokarst lake area changes in most cases is based on qualitative analysis of lake area changes. Such an approach makes it impossible to get complete quantitative data about lake area dynamics necessary for revealing regularities determining spatial and temporal changes of the fields of thermokarst lakes. In connection with this a necessity arises to do remote studying of lake fields, which should be focused on revealing the indicated regularities in the changing of these fields' properties.

So the main goal of this chapter of the monograph is presenting the results of quantitative study of thermokarst lakes dynamics in West siberian permafrost using space images.

1. Thermokarst Processes in the Permafrost

According to Mudrov (2007), permafrost is widely spread on earth and can be found practically on every continent. The area of its distribution is more than a quarter of the whole land including about 75 % of Alaska and more than half of Canada. In Russia the total permafrost area is 10.7 million km^2, which is about 63.5 % of the territory of the country (Geocryology of USSR, 1989).

Depending on the degree of permafrost density its continuous, discontinuous and sporadic distribution types are distinguished. In Russia most area (61.8 % of the total area of permafrost distribution) is taken by a continuous cryolite zone whose area includes most Arctic islands and stretches along nearly the whole coast of the Kara Sea in the west up to the Chukchi Sea in the east penetrating into the continent in Central Siberia and Yakutia (Geocryology of USSR, 1989). A sporadic cryolite zone (21.2 % of the total area of permafrost distribution in Russia) stretches from the Kola Peninsula to the coast of the Sea of Japan, Sakhalin and Kamchatka. The least area (17 % of the total permafrost area) is taken up a discontinuous cryolite zone situated between sporadic and continuous cryolite zones.

The West Siberian Plain occupying an area of more than 2 million km² and delineated to the west by the Ural Mountains, in the north – by the Kara Sea, in the east – by the Yenisei River, and in the south – by Kuznetsk Ala-Tau, the foothills of Altai and Kazakh Melkosopochnik is a unique area on the planet in terms of environmental conditions. The main distinctive feature of the Plain is its extremely high level of paludification caused by climatic and orographical features. Average peatland coverage of its territory is about 50 %, and in some areas such as Surgut Polesye, Vasyugan Plain, and the reservoirs of the rivers Ljamina, Pima, Anga it reaches 70-75 %. The huge number of lakes is another characteristic feature of the Plain. According to the data collected by the State Hydrological Institute, the general number of lakes in this territory exceeds 800,000. Taking into account all intra-mire water reservoirs like small lakes and ponds with area less than 1 ha, this number will considerably increase. The presence of a huge number of lakes among mires creates a specific mire-lake landscape on a significant part of the Plain territory (Mires of Western Siberia ..., 1976).

The West Siberian Plain is divided into two parts by the Siberian Ridge (Sibirskie Uvaly) which extends in an east-west direction. The southern part of the West Siberian Plain is characterized by landscape sub-zones, such as taiga and sub-taiga, forest steppes and steppes with different types of mires. Permafrost relief forms are only present in isolated frozen peat "islands" near slopes of the Siberian Ridge (Kirpotin et al., 2007). The most southern permafrost islands reaching 58° north latitude are very unstable and sensitive to climate-driven changes (Muldiyarov, 1987).

North of the Siberian Ridge the forest-tundra landscapes appear which gradually change into typical tundra in more northern regions. The forest-tundra zone is characterized by vast peatlands. The frozen peatlands form flat plateaus and large mound bogs. They constitute up to 70% of watershed surface. Lakes and fens are present over deeper permafrost locations or where the permafrost is absent, between flat mound bogs. Woodlands, open Larch (*Larix sibirica*), Fir (*Picea obovata*) and Pine (*Pinus sibirica*) forests cover up to 20% of the total surface. These species are found in the places where permafrost is absent, such as flood plains, terraces of large rivers, narrow strips zones around some lake shores, and on high mounds separating the bogs (Kirpotin et al., 2007).

Palsa bog (from Finnish – palsa) is poorly-drained lowland underlain by organic-rich sediments, which contains perennially frozen peat bodies: peat plateaux or peat mounds. Palsas are caused by permafrost and protrude above the surface of the surrounding peatland. The elevation height of these mounds may vary from less than one metre to several metres and may cover up to several hundreds of square metres in area whereas the larger palsas may form peat plateaux (Hoofgard, 2003).

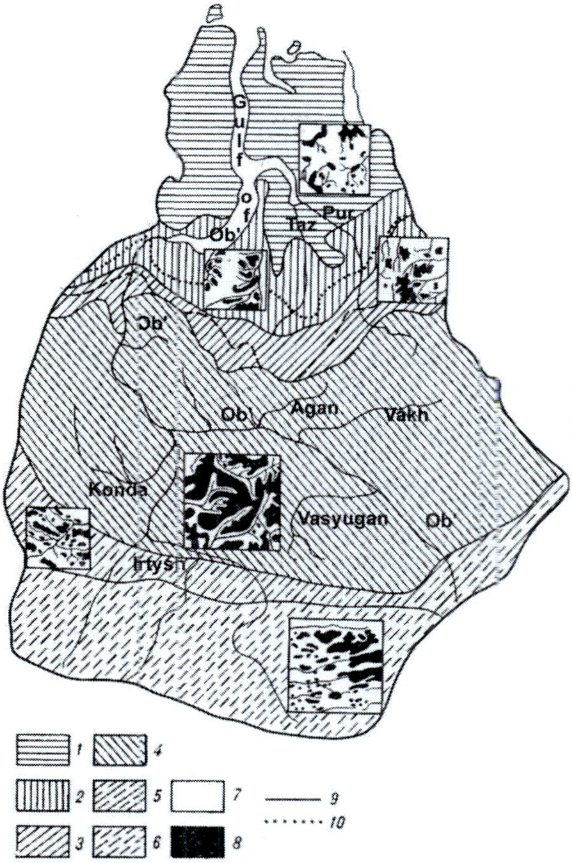

1 – zone of polygonal mires; 2 – zone of flat mound mires (palsas); 3 – zone of large mound mires (palsas); 4 – zone of raised oligotrophic sphagnum bogs; 5 – zone of flat eutrophic and mesotrophic sedge-hypnum mires; 6 – zone of concave eutrophic (reed) and salt grassy mires (salinas); 7 – uplands; 8 – mires; 9 – borders of the West Siberian Plain and mire zones; 10 – northern border of some large mound palsas.

Figure 1. Mire zones of the West Siberian Plain (from Mires of Western Siberia ..., 1976).

Palsa mires are mire complexes that occur in the northern hemisphere, representing one of the most marginal permafrost features at the outer limit of the permafrost zone [4] which makes them extremely sensitive to climatic fluctuations (Salmi, 1970; Seppala, 1988; Hoofgard, 2003; Luoto et al., 2004; Kirpotin et al., 2007). The upper temperature limit for palsa formation in Europe corresponds to the 1ºC mean annual temperature isotherm (Hoofgard, 2003; Seppala, 1988) and the southern limit coincides with the - 1 ºC mean annual air temperature (Salmi, 1970). In Western Siberia palsas exhibit wider distribution in comparison with Western Europe and North America being the dominant landscape type in the northern part of a northern taiga sub-zone, forest tundra sub-zone and southern tundra (Schumolova, 1962; Piavchenko, 1971; Tyrtikov , 1979; Malyasova et al., 1991; Novikov et al., 1999, Liss and Beresina, 1981; Muldiyarov, 1987; Vasiliev, 2007), occupying 280,000 km^2 (Mires of Western Siberia …, 1976) (Figure 1). In publications by Russian scientists palsas are called "mound bogs". Two main types of them are distinguished (Schumolova, 1962; Piavchenmko, 1971; Mires of Western Siberia …, 1976; Tyrtikov , 1979; Novikov et al., 1999; Liss and Beresina, 1981; Muldiyarov, 1987; Vasiliev, 2007): large mound palsas (with dome shaped tall mounds 2-5 (7) m high) ingrained into surrounding swamps and flat mound palsas (with vast plateaus low mounds (0,5-1,5 m)) corroded by numerous thermokarst subsidences of different shape. Large mound palsas occur in the southern band of mound bogs. They are typical for the southern part of a north-taiga sub-zone, shifting in the northern part of a north-taiga sub-zone and forest tundra sub-zone to the drained river valleys. Flat mound palsas occupy vast areas in the northern edge of a north-taiga sub-zone and forest-tundra, coming up to the southern tundra (Figure 1).

a

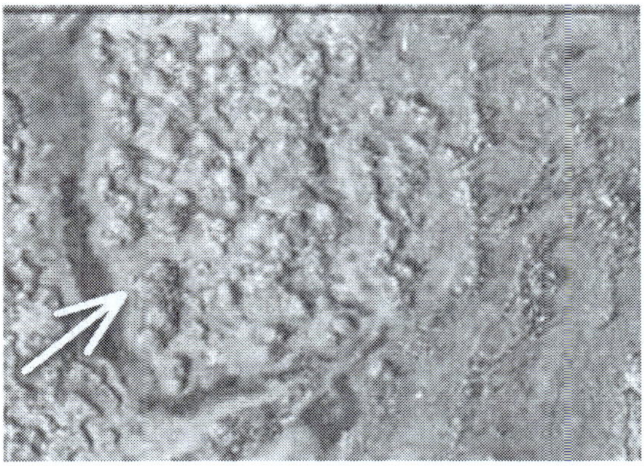

b

Figure 2. Large mound palsas: a – ground view (photo by S. Kirpotin, 2008), b – air photograph (19.07.1989).

The zone of flat-mound palsas is situated to the south of the polygonal bogs zone. The southern border of this zone is of a diffuse type and constitutes itself as a wide strip on which both flat-mound palsas and large-mound palsas are distributed. The area occupied by the zone of flat-mound palsas makes about 8 % (160 000 sq km) of the total area of the West Siberian Plain. The average water balance on the territory is the following: precipitation of 550 mm, runoff of 260 mm, and evaporation of 290 mm. The zone of large-mound palsas is located to the south of the flat-mound palsas zone (see Figure 2). According to Gorodkov (1935), the southern border of large-mound palsas coincides with the southern border of long-term frozen grounds. The area of this zone is rather insignificant and occupies only 6 % (120 000 sq km) of the total area of the West Siberian Plain. The precipitation, runoff and evaporation of the southern territory are equal to 600, 280 and 320 mm respectively (Mires of Western Siberia ..., 1976).

2. Thermokarst Lakes As the Most Suitable Indicator for Remote Study of Permafrost Changes

Thermokarst lakes are the most widespread landscape type in the northern part of north taiga, forest tundra and the southern edge of tundra.

Figure 3. Plenty of lakes and ponds in the Nadym-Pur watershed centre (photo by S. Kirpotin, 1999).

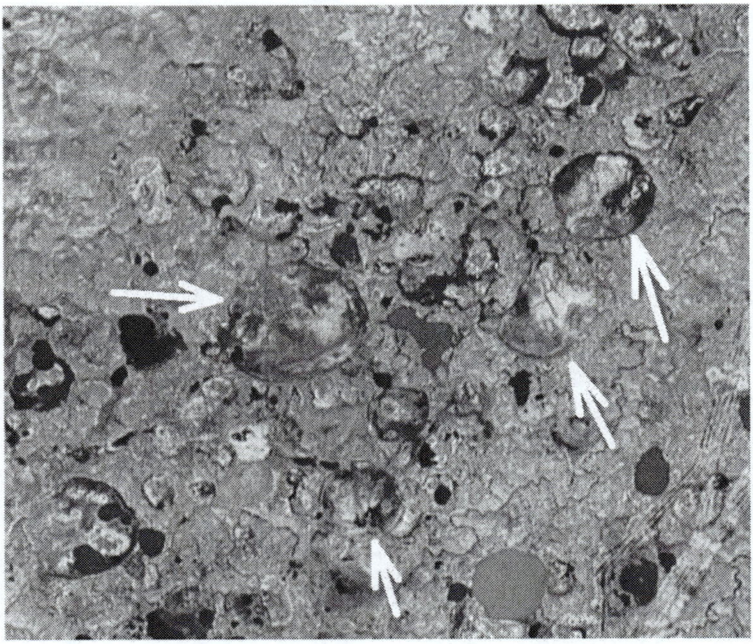

Figure 4. 'Khasyrei's field' in West siberian forest-tundra (from Google Earth).

The present-day surface of the frost flat peat mounds has attributes of sufficient degradation manifested by the prevalence of lichens in a vegetation cover and the presence of spots of bare peat (3-5%). Abundant presence of lakes (20-40% reaching up to 80% in the watershed centre) is typical for the frozen flat mound complexes of interstream areas. Figure 3 shows a typical field of thermokarst lakes in the central part of the West siberian permafrost

Approximately 70-95% of the surfaces free from lakes, with a thawed layer of 20-60 cm, are occupied by peat mounds of 0,5-1,5m height. The bog hollows accordingly occupy from 5% up to 30%. The permafrost is located at the 50-200 cm depth. The surface of mounds has a bulky-tussocks character: it is made by tussocks of the 20-30cm height, with the diameter of 0,5-2 m. Small shrub-lichen vegetation of the mounds is very typical. Small shrubs cover from 10% up to 30% of the land surface. The lichens (*Cladina stellaris* and *Cladina rangiferina*) usually occupy 70-80% of the surface. The mosses are found in separate spots and occupy no more than 2-5% of the area.

Drained thaw lake fields in Western Siberia have a local name "khasyreis" (Kirpotin et al., 2007). Khasyreis as well as lakes, ponds and frozen mounds are invariable element of sub-arctic frozen peat bogs – palsas and tundra landscapes. In some areas of West siberian sub-arctic khasyreis occupy up to 40-50 % of total lake area. Sometimes its concentration is so high that we call such places 'khasyrei's fields" (Figure 4). Khasyreis are part of the natural cycle of palsa complex development (Kirpotin et al., 2007), but their origin is not continuous and uniform in time and, according in our opinion, there were periods of more intensive lake drainage and khasyrei development accordingly. These times were correspond to the epochs of climate warming and today we are faced with one of them.

In remote studies of dynamics of thermokarst lake fields it is important to choose proper geoindicators of cryogenic changes. The analysis of using remote sensing data in geocryological research showed (Melnikov et al., 1974; Gudilin and Komarov, 1978; Polishchuk and Tokareva, 2006) that thermokarst lakes which are easy to decode in satellite images are the most suitable indicators of geocryogenic changes in the surface under the conditions of permafrost landscapes in northern territories. An example of a satellite image of a field with thermokarst lakes and khasyreis is presented in Figure 5.

Fields with thermokarst lakes display noticeable changes in both time and space: new lakes emerge while some lakes disappear turning into khasyreis (hollows of drained lakes); the shape of lakes' boundaries and areas of their water level change. Let us consider cyclic processes of thermokarst lakes' evolution in permafrost landscapes. Figure 6 is an illustration showing a succession of stages of cyclic changing thermokarst landscape in the permafrost zone.

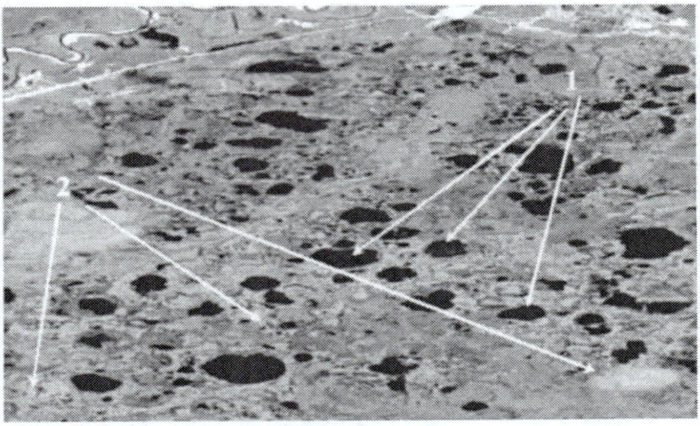

Figure 5. Space image Landsat-7 (07.08.1999) with thermokarst lakes (1) and khasyreis (2).

a

b

c

d

Figure 6. Stages of cyclic succession of palsa complex development: a – thermokarst subsidence; b – initial thermokarst lake; c – young thermokarst lake; d – khasyrei (photo by S. Kirpotin, 2005).

The frost flat mound bogs, as a rule, have an insignificant peat deposit of mixed type with the thickness of 0.5-1.5 m, in rare cases reaching 3-5 m, combined mainly by fen-sedge, sedge-equisetum, sedge-hypnum-moss and hypnum-moss types of peat; while the peat deposits of the bottom layer of bog hollows are close in botanical composition to deposits of the frost peat mounds and are found to have less thickness (Schumilova, 1962; Tyrtikov, 1979; Novikov and Usova, 1983). It testifies the secondary origin of bog hollows, beginning after cessation of active peat accumulation process in the optimum of Holocene, when the body of peat bogs developed as a single unit (Tyrtikov, 1979; Novikov and Usova, 1983). The absolute age of the bottom layer of peat deposit of mounds is from 5280 (+ – 240) to 9320 (+ – 120) years old, and of the bog hollows - from 2860 (+ – 90) to 4750 (+ – 110) years old (Novikov and Usova, 1983).

The information on joined pollen and botanical analysis of peat, available in publications, allow restoring the stages of evolution of the frost flat mound bogs (Novikov and Usova, 1983). The accumulation of basic masses of peat proceeded in the period of warmer postglacial time in conditions of deep permafrost deposition, when the territory under consideration was probably similar to a northern taiga subzone. In this period the fen-senge-hypnum-moss with equisetum, comarum and menyanthes types of peat were mainly formed. The formation of the dry and cold climate at the next stage resulted in occurrence of permafrost in the peat bogs, which was shown in cryogenic heaving of the surface of low-level bogs. The existence of two phases of the permafrost accumulation is evidenced from the data of geomorphology of double-stage structure of the permafrost rocks (Zemtsov, 1976). The next climate warming in the middle of Holocene, proceeding until now, has initiated processes of permafrost degradation, formation of inter-mound hollows and their swamping (Novikov and Usova, 1983).

There are three basic hypotheses on the origin and genesis of palsas in publications by Russian scientists. According to the first hypothesis (Dranitsin, 1914; Popov, 1945), the "frozen heaving" of mounds, negative forms of a relief – bog hollows - were formed primarily. According to these authors, mounds have arisen as a result of freezing of initially thawed mires (Kirpotin et al., 2003).

The second hypothesis was formulated by Piavchenko (1971). According to him, positive forms of relief in mound bog complexes represent frozen relics peatbogs, and bog hollows are new formations appeared owing to destruction of permafrost and subsidence of mire ground (Kirpotin et al., 2003).

The third hypothesis, which has been put forward by Shumilova (1962), is more compromise, and suggests the dual genesis of the mound's relief, connected both with permafrost heaving and thermokarst.

In recent publications we have described the endogenous cyclic succession of palsa complex development (Kirpotin et al., 2007; Kirpotin et al., 2008). In the first stage of this succession the flat mounds through series of subsidences and hollows of different degrees of moistening turn into a thermokarst lake. In the second stage, the lake inevitably turns into a khasyrei.

The most typical way of khasyrei origin is their drainage to bigger lakes which are always situated on the lower levels and act as a collecting funnels providing drainage of smaller lakes (Figure 7). The lower level of a big lake appears when the lake accumulates a critical mass of water enough for subsidence of the lake bottom due to the thawing of underlying frozen rocks (Kirpotin et al., 2008).

Figure 7. Sample of a typical big lake surrounded by small drained lakes (photo by S. Kirpotin, 1990).

Figure 8. Khasyrei drained by small Puritey-River (air photograph 19.07.1998).

Another way of lake drainage is the lake intercept by a river (Figure 8). Some authors believe (Hinkel et al., 2003; Smith et al., 2005; Romanovsky et al., 2007) that the most typical way of khasyrei origin is the lake drainage to the subsurface (underlying rocks). According to our opinion, it is not possible in Western Siberia, where the thickness of the permafrost is 300-500 m constituting safe confining bed. In addition, peat of Western-Siberia presents an excellent insulating layer further protecting the frozen rocks from thawing. Even under the lake bottom. Extremely shallow depth of West siberian lakes leads to their full freezing in winter also decrease the probability of drainage in the subsurface horizons.

Thermokarst lakes are one of the most climate-sensitive elements of permafrost landscapes. They are used as a suitable indicator of climatic changes. Furthermore, thermokarst lakes have effect on climate because they are an important source of CO_2 and methane emission into the atmosphere. Ongoing processes of the permafrost thawing in Western Siberia are likely to increase the surface of water reservoirs via forming the so-called thermokarst lakes, mobilizing the organic carbon (OC) from the soil pool to the rivers and, finally, to the ocean, and thus modifying the fluxes of methane and CO_2 to the atmosphere (Walter et al., 2006). The overall potential of carbon loss from yedoma (permafrost) soils under scenario of continuing warming is estimated as 1 Gt C yr-1 or ~40 Gt C in the next four decades (Dutta et al., 2006). Significant part of this carbon flux will be mediated by active hydrological systems such as thaw lakes. Indeed, it is known for long that the boreal lakes yield net annual CO2 evasion to the atmosphere due to the dominance of net heterotrophy (Cole et al., 1994; Hope et al., 1996; Kelly et al., 2001; Sobek et al., 2003). Up to present time, however, all measurements of this flux dealt used direct (chambers) or indirect (isotopes, pCO_2 values) techniques to assess the concentrations of CO_2 in the boundary air or in the liquid phase (e.g., Kling et al., 1991) whereas the chemical and microbiological mechanisms of this process remained poorly understood. It can be suggested, that, by analogy with non-permafrost affected environments (Tranvik, 1988; Kritzberg et al., 2004), the net heterotrophic status of thermokarst lakes can be due to bacterial utilization of allochtonous organic carbon originated from the peat erosion. (Shirokova et al. 2013). Since all lakes on the permafrost ground have thermokarst origin and thus are similar to those studied in this work, one calculate the minimal flux of the CO_2 to the atmosphere from 792,000 km^2 of the whole permafrost lake area, showing that 65.5% of lakes north of the Arctic Circle occur in Russia (Walter et al., 2007). The annual input of CO2 from the earth surface to the atmosphere due to thermokarst lakes presence thus achieve 0.1 Gt C/y (Shirokova et al. 2009) which is significantly higher than the total riverine flux of dissolved and suspended organic carbon from

Russia territory to the Arctic ocean (0.02 Gt C/y, Romankevitch and Vetrov, 2001). Walter et al., (2007) estimated lake area of Western Siberia using the fine-scale lake database of Lehner and Doll (2004), which is the best one available. But they excluded lakes smaller than 0.1 km^2 and therefore significantly underestimated thermokarst lakes number and area (Frey and Smith 2007; Walter et al., 2007). At the same time, as Grosse et al. (2005) have shown, these small lakes are particularly important contributors to CH_4 ebullition because they are the most numerous lakes in North Siberia and, on the other hand, they "have the largest fluxes per unit area because their low area: perimeter ratio causes lake-margin carbon inputs (thermokarst erosion and aquatic plant production) to be relatively more influential" (Walter et al., 2007).

3. Methodical Problems of Remote Research of Thermokarst Lake Dynamics

Figure 9 presents a fragment of a satellite image showing a set of thermokarst lakes of different size. Studying dynamics of state of thermokast lake field involves studying changes in thermokarst lakes' areas. To do this, areas of lakes' water levels were measured in non-simultaneous satellite images using geoinformation systems (GIS) software. This is the reason why it is important to detect reliably thermokarst lakes' coastal frontiers while decoding satellite images.

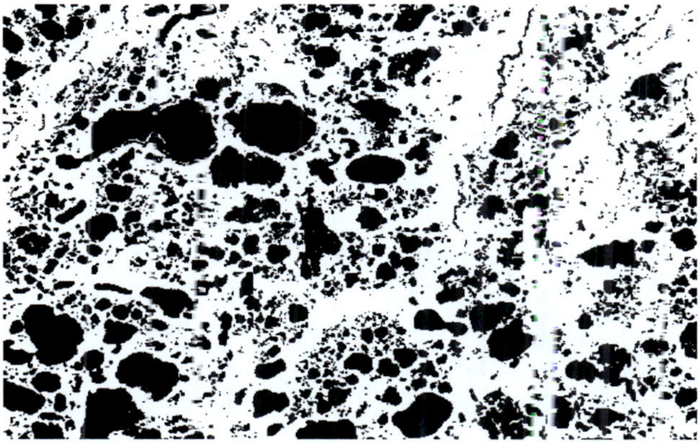

Figure 9. Fragment of a satellite image from Landsat-5 (13.07.2007) with thermokarst lakes.

To determine quantitative characteristics of dynamics of lakes' areas having been observed for many years it is important to precisely measure lakes' areas with the help of satellite images. In a general case the accuracy of determination of characteristics for observed for a long time thermokarst lakes' dynamics is affected by both measurement bias while measuring thermokarst lakes' areas in satellite images and intra-seasonal changes in thermokarst lakes' areas caused by floods, rain, snow thawing, etc.

Biases of remote measuring lakes' areas and the influence of intra-seasonal thermokarst dynamics on changes in lakes' areas are studied in the paper by Bryksina and Polishchuk (2013). They suggest a technique to choose test sites intended to studying biases when measuring lakes' areas in optical and radar images. The results of studying accuracy of lakes' areas measurement showed that for the lakes with the area of 2-4 ha medium relative bias does not exceed 3-5 %, and for the lakes with the area of more than 20 ha it is 1–2 %. In their research Bryksina and Polishchuk (2013) use radar images from ERS-2 and optical images of ultrahigh spatial resolution Quick Bird.

The technique of thermokarst dynamics study requires measurement of lake areas by means of non-simultaneous space images and accuracy of results depends on magnitude of intra-seasonal changes of lake area. Intra-seasonal dynamics of thermokarst lake areas were studied. Because of a small number of cloudless days in these northern regions, only radar space images could be used. Complete methodology and results of this research using radar satellite images from ERS-2 and ENVISAT are given in Bryksina and Polishchuk (2009).

To study intra-seasonal lake area changes in the West Siberian permafrost 6 test sites were chosen (Bryksina and Polishchuk, 2009), which are situated in different permafrost subzones. For each site radar images from Landsat-5 and Landsat-7 were selected within a warm period of a year from May to September. The whole number of images in use for a two-year period was 42, including 8 images from Landsat, 30 images from ERS-2 and 4 images from ENVISAT.

Figure 10 shows a diagram of lake area changes during warm seasons of 2005 and 2006. Averaged areas for test sites in discontinuous permafrost are marked by dots. Confidence intervals for 0.95 are shown by cuts. As diagrams indicate, mean deviation of lake areas is negligible in all test sites during warm season in every year. Simple mean of relative deviation of lakes areas for test sites in different permafrost zones is less than 1 - 2%. It shows small influence of intra-seasonal relative deviation of lake areas on the results of measurement of lake areas for research of multiannual areas dynamics based on optical images from Landsat chosen during 1973-2011.

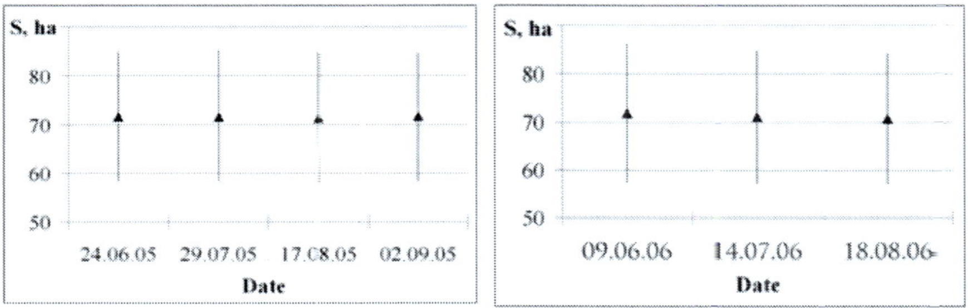

Figure 10. Time dependences of thermokarst lakes areas in a discontinuous permafrost.

Research into the multi-year dynamics of thermokarst lakes in the West siberian north was carried out by means of measurement of lakes' areas using the multi-year collection of satellite optical images from Landsat taken at different times in the above 30-year period. Areas of thermokarst lakes on all the test sites were measured basing of satellite images. The total number of lakes was from several hundreds to several thousands. Measurements of lake areas were carried out using geoinformation systems ERDAS Imagine 8.3, ENVI 4.4 and ArcGIS 9.3. The array of measured data have been analyzed statistically for research in thermokarst changes.

The study used data on the areas (sizes) of the thermokarst lakes produced using satellite imagery for 30 test sites (TS), located at places of active thermokarst in the permafrost territory of Western Siberia. Figure 11 (left) represents a map-scheme of geocryological zoning of the West siberian permafrost, which shows the location of studied test sites in zones of sporadic, discontinuous and continuous permafrost. A scheme of landscape zoning of the West siberian permafrost is given in Figure 11 (right), which shows the location of these test sites in landscape zones of the permafrost territory in Western Siberia. As it is seen in Figure 11, test sites were chosen so as to more evenly cover the territory of research.

It is worth mentioning that the zones of continuous and discontinuous permafrost are situated in Yamalo-Nenets Autonomous District, and the sporadic permafrost zone – in the central part of West Siberia, mainly in Khanty-Mansi Autonomous District and the northern part of Tomsk Oblast.

An important issue for studies of the dynamics of lakes is compiling a collection of satellite images. The satellite optical images from Landsat were chosen for research as there is multi-year archive of these images. The technique of thermokarst dynamics study supposes measurement of the lakes' areas by means of the non-simultaneous space images. However, the accuracy of results of thermokarst dynamics research depends on magnitude of intra-seasonal changes of lake areas. Intra-seasonal dynamics of thermokarst lake areas was studied in our work. Because of a small number of cloudless days in the northern territories only radar space images may be used for intra-seasonal dynamics research of thermokarst lakes. Methodology and results of this research using radar satellite images ERS-2 are given in Bryksina and Polishchuk (2009). This research has shown that intra-seasonal changes in thermokarst lakes areas on different test sites are on average less than 1 - 2 %. This permits the collection of Landsat satellite images to be augmented by choosing cloudless optical images obtained in different months during the warm season.

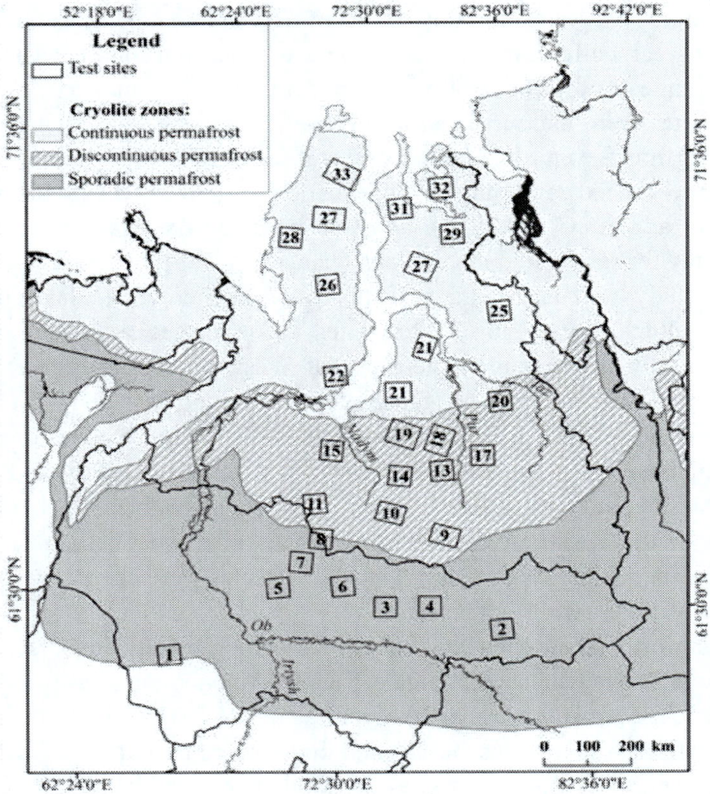

a

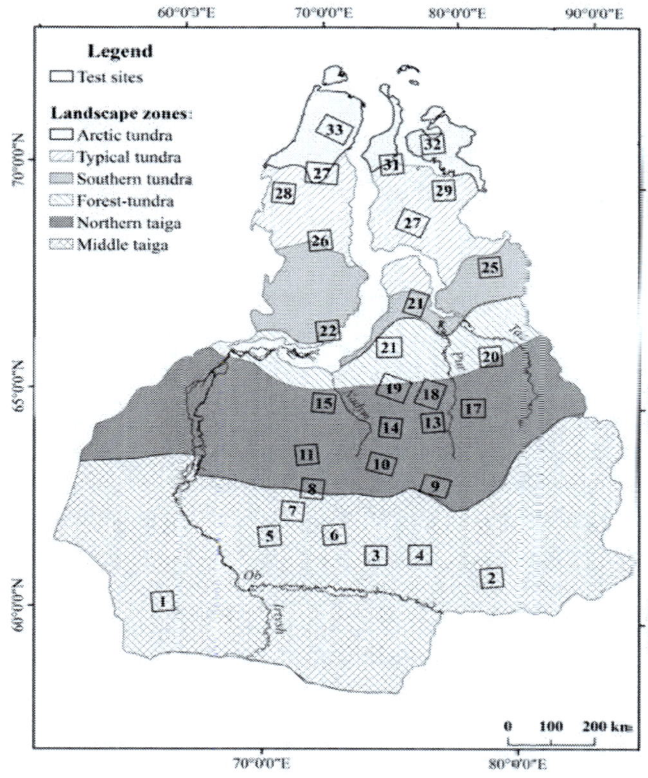

b

Figure 11. Geocryological map and landscape map of West siberian lowlands. Distribution of sporadic, discontinuous and continuous permafrost across the West siberian lowlands and locations of test sites (a). Distribution of types of regional landscapes and test sites (b). The locations of test sites were chosen to evenly represent different landscapes and permafrost types.

As a result, the collection formed included 106 cloudless space images received during the period of 1973 – 2009 selected from the USGS Global Visualization (http://glovis.usgs.gov):

1) Landsat - 1 (scanner MSS, spatial resolution 80 m), 1973;
2) Landsat - 2, 4, 5 (scanners MSS and TM, spatial resolution 30 m), 1981-2009;
3) Landsat - 7 (scanner ETM+, spatial resolution 30 m), 1999-2003

The areas of thermokarst lakes on all test sites were measured basing on these satellite images and the total number of lakes in each test site ranged from several hundreds to several thousands lakes. Figure 3, 4 and 5 show a typical fields of thermokarst lakes in the central West siberian permafrost. The range of measured data has been analyzed statistically for understanding changes in thermokarst lake fields.

4. The Analysis of Trends in Changing Thermokarst Lakes' Number and Area

Thermokarst lake fields suffer from temporal and spatial changes including changes in lake area, lakes' disappearing (forming khasyreis) and formation of new lakes. Theses processes have accelerated under the conditions of contemporary climate warming. Thermokarst lake area is one of the most sensible indicators. To illustrate this, Figure 12 displays fragments of non-simultaneous satellite images presenting successive stages of changes in a thermokarst lake's area (Dneprovskaya et al., 2009). The area of a lake's water surface was measured with the help of four satellite images taken at different time, whose fragments are shown in Figure 12.

Lake area measurement showed that in1993 the area of its water surface (Figure 12-b) decreased to 60 % from initial value of the lake's area measured in 1973 (Figure 12-a). The process of decrease in the lake's area is still going on, which is confirmed by the images taken in 2002 and 2005 (Figure 12-c and 12-d). As the latest image displays, for a 30-year period of time the lake's area decreased by 2005 to 40 % of its initial value. To illustrate this, Figure 13 shows fragments of non-simultaneous satellite images presenting successive stages of decrease in the lake's area and its turning into a khasyrei (Figure 13-d) (Dneprovskaya et al., 2009). The lake's area was measured with the help of four non-simultaneous satellite images.

Measurements done in 1988 showed (Dneprovskaya et al., 2009) that the lake's area decreased (Figure 13-b) nearly to 20 % of initial value measured in 1973. In 1993 the lake's area decreased to 3 % of its initial value, and in 2005 the lake disappeared completely having turned into a khasyrei. In Figure 13-d one can clearly see the shape of the basin of the extinct lake, which is notable for its darker colour in a panchromatic image by Spot-5. The thermokarst lakes in Figure 12 and 13 are the examples of the lakes whose water surface areas decreased very fast. Alongside with such lakes one can also observe thermokarst lakes whose areas increased or did not change for the period under study.

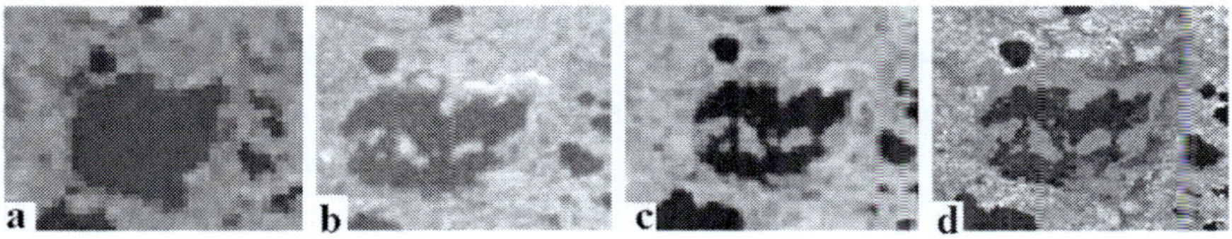

Figure 12. Fragments of satellite images showing the succession of stages of the lake's area changing: a – Landsat-4 (10.08.1973); b – Resurs-F2 (14.06.1993); c – Landsat-7 (03.07.2002); d – Spot-5 (20.07.2005).

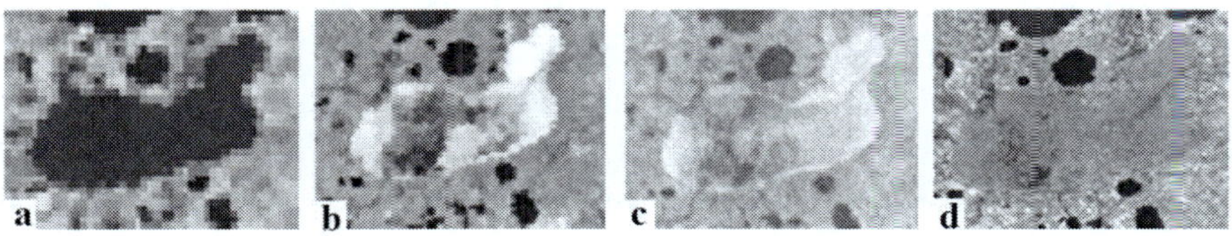

Figure 13. Fragments of satellite images showing the final stage of basin formation: a – Landsat-4 (10.08.1973); b – Landsat-5 (26.06.1988); c – Resurs-F2 (14.06.1993); d – Spot-5 (20.07.2005).

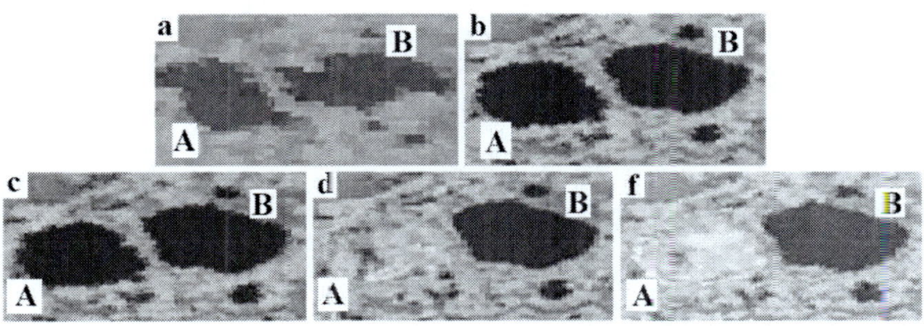

Figure 14. Fragments of satellite images showing the successive stages of changing areas of lakes A and B in TS-13: a – Landsat-5 (1984); b – Landsat-4 (1988); c – Landsat-5 (1989); d – Landsat-7 (1999); f – Landsat-7 (2001).

Figure 14 presents fragments of non-simultaneous satellite images showing successive stages of khasyrei formation in the place of an extinct thermokarst lake A (Dneprovskaya et al., 2009). Successive measurements of the water surface area of lake A done with the help of non-simultaneous images showed that by 2001 the water surface area of the lake (Figure 14-f) went down by 90 % in comparison with the initial value measured in 1984 (Figure 14-a). At the same time the area of neighbouring lake B (Figure 14) changed insignificantly for the period under study.

Owing to increased thermokarst activity, two contrasting processes are observed in the West siberian cryolithozone: increase in lake surface due to thawing of lakes' coasts (mainly in Northern part of it): and decrease in surface area or disappearance of lakes due to water escape downstream the bigger lakes and hydrological network (Kirpotin et al., 2008). Illustrations of both processes of increase /decrease in lake surface areas and disappearance of lakes have been given in (Bryksina et al., 2007; Samsonov et al., 2007; Kirpotin et al., 2008) on base of comparison of space images received in different times.

From the point of view of studying thermokarst lake area dynamics it is interesting to consider time dependence of changes in average values of thermokarst lakes' areas in TS-13. In connection with this 2 sets from the sample of lakes under study in TS-13 were formed (Dneprovskaya et al., 2009), the first of which included the lakes whose areas decreased, the second included the lakes whose areas increased. For each set average values of lake areas (for every year under study separately) were determined. In Figure 15-a there is a diagram showing the dependence of changes in average area (S) of a set of lakes whose areas decreased on time (T). A similar diagram of time dependence on average area of a set of the lakes whose areas increased is shown in Figure 15-b. Confidence intervals for average areas shown in Figure 15 have been calculated for the probability of 90 %.

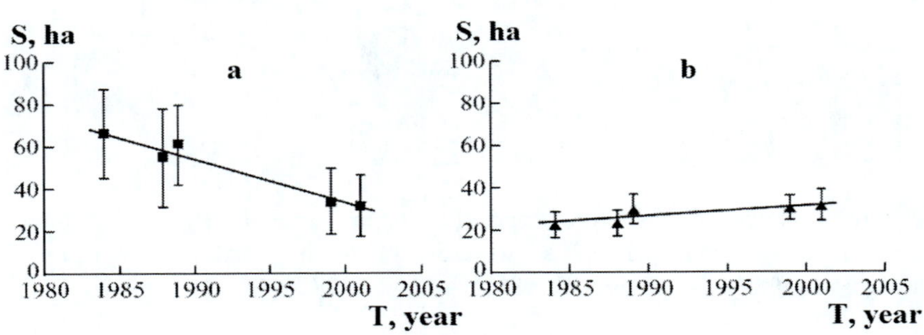

Figure 15. Dependence of average area of lakes on time for sets of lakes with reduced (a) and increased (b) water surface area.

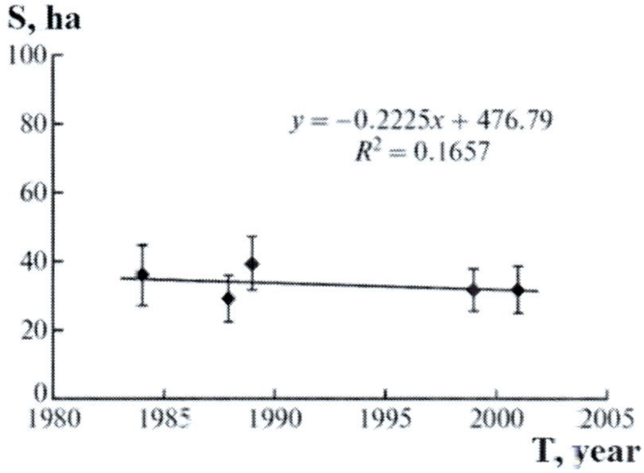

Figure 16. Dynamics of average area of lakes on TS-13.

Comparing the diagrams in Figure 15-a and 15-b shows that on average the process of lakes' water surface area reduction prevails over the process of its increase. This conclusion is confirmed by results shown the diagram in Figure 16 which presents time dependence on the changes in average area of all the lakes under study in TS-13 showing decrease in the slope of the trend line compared to the diagram of time dependence on average area in Figure 15-a. Consequently, as it can be seen from the diagram in Figure 16, thermokarst lakes' areas during the observation period in the territory with center coordinates of 75° east longitude, 66° north latitude on average reduce in time, which reveals an important trend in contemporary dynamics of thermokarst processes in the West Siberian discontinuous permafrost (Dneprovskaya et al., 2009).

Possible reasons for changes in lakes' areas have been considered above. There are two most probable mechanisms accounting for lake area reduction: 1) soil drainage under permafrost thawing because of soil temperature rise caused by climate warming; 2) transpiration by vegetation, whose intensity rises together with the increase in air temperature.

The number and density of distribution of thermokarst lakes over the territory of test sites (TS) have been studied. Comparison of the data on density of lakes' distribution over different TS shows that in the West Siberian permafrost in the period of 1973-2011 average values of lakes' density changed considerably for each TS (from 0,15 to 0,92 lakes per 1 km^2) and on average was 0,41 lakes per 1 km^2 of TS area, and the number of lakes in each TS changed from 389 to 2731 and is, on average, 1129 lakes.

5. Analysis of Results of Multi-Year Thermokarst Dynamics Research

In order to study thermokarst dynamics depending on geographical latitude and landscape type summarized (total) areas of thermokarst lakes were determined in each test site and for the each year of observation. Value of relative changes of summarized areas is calculated by formula:

$$R = (S_f - S_i) / S_i,$$

where S_i and S_f - summarized areas of lakes on initial and final year of research period accordingly.

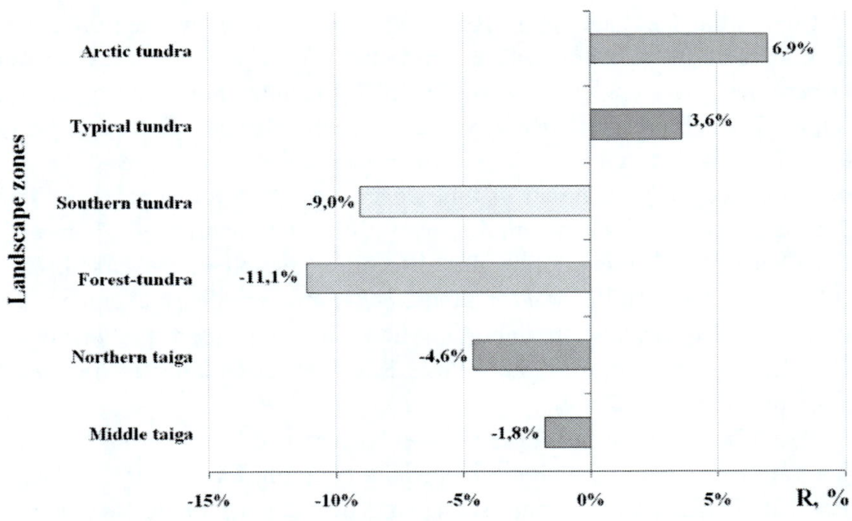

Figure 17. The average value of the relative change lake total area in different landscape zones.

Figure 17 represents a diagram of the relative change in the total area of lakes (R_m) for the period of study, which is calculated in Bryksina et al. (2012) as the arithmetic mean of the individual values of R for the test sites located in selected landscape zones. We takes into account the lakes area of 1 hectare or more, which are available on the Landsat images. According to Figure 17, increasing the total area of lakes is only in two zones of the tundra - the Arctic and typical ones. In zones of the southern tundra, forest-tundra, northern and middle tayga in which Rm <0 a tendency is observed to reduce the total area of the lakes. It should be noted that in areas of the forest-tundra and southern tundra tendency to reduce the total area of lakes are extreme. Thus, the values of relative change in the total area of the lakes in these two zones are -11.1% and 9% respectively.

It is interesting to consider the thermokarst dynamics depending on the latitude. This study used time series of the total area of lakes in the test sites in different years of the study. Linear trends of time series were determined in accordance with the equation of the linear approximation:

$$y = \alpha x + \beta, \quad (1)$$

where α - linear coefficient of approximation equation;

β - constant term of approximation equation.

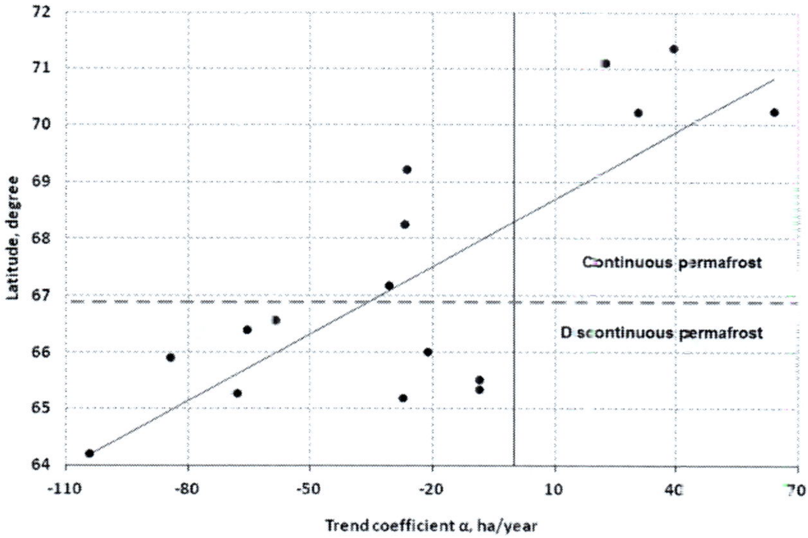

Figure 18. Liner trend coefficient for total area of lakes depending on latitude.

The positive sign of the coefficient α indicates that the analyzed indicator (in this case, the area lakes) in the analyzed period displayed the average increase, while a negative value ("-" sign) of coefficient α indicates an average reduction rate in the analyzed period of observations. A graph of changes of the coefficient α depending on the latitude is shown in Figure 18. The solid line on the chart, which displays an approximating straight line, shows the trend of growth in values of the coefficient α with increasing latitude of test site location. This is a manifestation of the growth trends of the total area of thermokarst lakes in the West siberian permafrost in the direction from south to north.

As seen in Figure 18, increase in the total area of lakes is in northern part of the continuous permafrost zone (test sites 27–30 in Figure 17 located above 70 deg. N). Other test sites located south of the 70th latitude both in the southern part of the continuous permafrost and in the discontinuous zone reveal the tendency of decrease in the total area of lakes. This can be interpreted as the displacement of the modern boundary between continuous and discontinuous permafrost to the north under the influence of global warming.

CONCLUSION

Thermokarst lakes are considered as the most suitable indicator of climatic changes in permafrost landscapes on base of remote sensing. Methodological issues for evaluating the accuracy of remote measurement of lakes areas based on optical and radar images were developed. The intra-seasonal dynamics of thermokarst lakes areas was studied using radar images ERS-2 and ENVISAT. It is shown that the intra-seasonal variation in the areas of thermokarst lakes in average do not exceed 2%. Studies of long-term dynamics of thermokarst on the base of space image from Landsat were carried out at 30 test sites located in different zones of permafrost and in different landscape zones. It is shown that the total area of lakes increases in a typical sub-Arctic tundra up 7% and it reduces in areas of the south tundra, forest-tundra and northern taiga up to 11%. It is found that of the quantity of newly formed lakes in the tundra increased to 40%, and in the area of forest-tundra and northern and central taiga - up to 15%, which can be associated with thawing of permafrost due to global warming. Remote studies of long-term dynamics of thermokarst lake areas have revealed the growth trends of the total area of thermokarst lakes in the West siberian permafrost in the direction from south to north. It is established that the increase in the total area of lakes observed in the northern part of the continuos permafrost above 70 degrees. In the southern part of continuous permafrost (south of the 70th latitude), a tendency to reduce the total area of thermokarst lakes is observed. This can be interpreted as the displacement of the current border between the continuous and discontinuous permafrost to the north under the influence of global warming.

ACKNOWLEDGMENTS

This research was financially supported by EU-INTAS project 34.35.25, CAR-WET-SIB "Biogeochemical cycle of carbon in wetlands of Western Siberia" GDRI (Groupement de recherche international) project, ANR project "IMPACT-Boreal", and RFBR 08-04-92495-CNRS_a, RFBR 08-05-92496-CNRS_a, RFBR 08-05-92497-CNRS_a projects. Landscape studies in August 2006 were carried out with the assistance of US National Public Radio. The remote results have been obtained in frames of Russian Federal Programme "Kadry" (contract No 14.740.11.0409), ESA CAT-1 Project ID:5762 "Cryogenic processes of monitoring in Russian permafrost territories using radar and optical space images" and the Megagrant of Russian Federal Programme 14.B25.310001 (BIO-GEO-CLIM).

REFERENCES

Bryksina, N.A., Yevtyushkin, A.V., Polishchuk, Yu.M., (2007). Study of dynamics of thermokarst relief forms using space images // *Modern problems of remote sensing of Earth from Space* (Ed. by E. Loupyan), Vol. 4, 2. Moscow. Izd. "Azbuka-2000". 123-128 (in Russian).

Bryksina, N.A., Polishchuk, Y.M. (2009). Analysis of seasonal changes of thermokarst lakes areas in permafrost zone of Western Siberia using satellite images ERS-2. *Research of Earth from Space*. 3, 90-93 (in Russian).

Bryksina, N. A., Polishchuk,V. Yu., Polishchuk, Yu. M., (2011). Database of thermokarst lakes in Western Siberia on the basis of satellite images and the possibility of its practical use. *Modern problems of remote sensing of the Earth from Space*. 8(3), 175-181. (in Russian).

Bryksina, N. A., Polishchuk, Yu. M., Bulatov, V. I. , (2012). Landscape-space analysis of the dynamics of fields of thermokarst lakes in the permafrost zone of Western Siberia. *Bulletin of the Tomsk State Pedagogical University*. 7, 167-170. (in Russian).

Bryksina, N. A., Polishchuk, Yu. M. (2013).Investigating the accuracy of remote sensing area lakes using satellite images. *Geoinformatics*. 1, 52-68. (in Russian).

Casper, P., Maberly, S.C., Grahame, H., Hall, G.H., Finlay, B.J. (2000). Fluxes of methane and carbon dioxide from a small productive lake to the atmosphere. *Biogeochemistry*. 49, 1–19.

Cole, J.J., Caraco, N.F., Kling, G.W., Kratz, T.K., (1994), Carbon dioxide supersaturation in the surface waters of lakes. *Science* 265. 1568-1570.

Dneprovskaya, V. P., Bryksina, N. A., Polishchuk, Yu. M. (2009). The study of thermokarst changes in the area of discontinuous permafrost in Western Siberia on the base of satellite images. *Study of Earth from Space*. 4, 88-96. (in Russian)

Dranitsyn, D.A. 1914. About some zone forms of a relief of Far North. *Soil science* 4 (in Russian).

Dutta, K., Schuur, E.A.G., Neff, J.C., Zimov, S.A. (2006) Potential carbon release from permafrost soils of Northeastern Siberia. *Global Change Biology* 12, 2336–2351, doi: 10.1111/j.1365-2486.2006.01259.

Frey, E. F., Smith, L. C. (2007) How well do we know northern land cover? Comparison of four global vegetation and land cover products with new ground-truth database for West Siberia. *Global Biogeochem. Cycles* 21, GB1016. (doi:10.1029/2006GB002706).

Geocryology of USSR. (1989). Western Siberia / Ed. E.D.Ershov, 454.

Gorodkov, B.N. (1935). Vegetation of tundra zone of the USSR. *Vegetation of the USSR*. Moscow-Leningrad. 297-354 (in Russian).

Grosse, G., Shirrmeister, L., Kunitsky, V. V. & Hubberten, H.-W. 2005. The use of CORONA images in remote sensing of periglacial geomorphology: an illustration from the NE Siberian Coast. *Permafrost Periglac. Process* 16, 163-172. (doi:10.1002/ppp.509).

Hamilton, J. D., C. A. Kelly, J. W. M. Rudd, R. H. Hesslein, and N. T. Roulet (1994), Flux to the atmosphere of CH4 and CO2 from wetland ponds on the Hudson Bay lowlands (HBLs). *J. Geophys. Res.*, 99(D1), 1495–1510.

Hinkel, K.M., Eisner, W.R., Bockheim, J.G., Nelson, F.E., Peterson, K.M., and Dai, X. (2003). Spatial Extent, Age, and Carbon Stoks in Drained Thaw Lake Basins on the Barrow Peninsula. Alaska. *Arctic, Antarctic, and Alpine Research*. 35(3), 291-300.

Hoofgard, A. (2003). Effects of climate change on the distribution and development pf palsa peatlands: background and suggestions for a national monitoring project. *NINA Project Report 21*. Trondheim. Norway, 33.

Hope, D., Kratz, T.K., Riera, J.L., (1996), The relationship between pCO2 and dissolved organic carbon in the surface waters of 27 northern Wisconsin lakes. *J. Environ. Qual*. 49, 1442-1445.

Kelly, C.A., Fee, E., Ramlal, P.S., Rudd, J.W.M., Hesslein, R.H., Anema, C., Schindler, E.U., (2001), Natural variability of carbon dioxide and net epilimnetic production in the surface waters of boreal lakes of different sizes. *Limnol. Oceanogr.* 46, 1054-1064.

Kirpotin, S.N., Blyakharchuk, T.A., Vorobyev, S.N. (2003). Dynamics of Subarctic West Siberian Palsas as Indicator of Global Climatic Changes. Bulletin of Tomsk State University. Appendix 7. Proceedings of the Trance-Regional Ecological Workshop "Comprehensive Ecological Exploration of Siberian Landscapes" (September 2003). Tomsk: TSU, 122-134 (In Russian).

Kirpotin, S.N., Naumov, A.V., Vorobiov, S.N., Mironycheva-Tokareva, N.P., Kosych, N.P., Lapshina, E.D., Marquand, J., Kulizhski, S.P., Bleuten, W. (2007). Western-Siberian Peatlands: Indicators of Climate Change and Their Role in Global Carbon Balance. Chapter 33 in Climate Change and Terrestrial Carbon Sequestration in Central Asia / edited by R.Lal, M.Suleimenov, B.A.Stewart, D.O.Hansen, and P.Doraiswamy, Taylor and Francis, Amsterdam, Holland, 453-472.

Kirpotin, S.N., Polishchuk, Yu. M., Bryksina, N.A. (2008). Dynamics of thermokarst lakes areas in continuous and discontinuous cryolithozones of Western Siberia under global warming. *Vestnik of Tomsk State University*, 311. 185-189 (in Russian).

Kirpotin, S., Polishchuk, Yu., Zakharova, E., Shirokova, L., Pokrovsky, O., Kolmakova, M., Dupre, B. (2008). One of Possible Mechanisms of Thermokarst Lakes Drainage in West siberian North. *International Journal of Environmental Studies*. 65(5), October 2008, 631-635.

Kirpotin, S., Polishchuk, Y., Bryksina, N. (2009). Abrupt changes of thermokarst lakes in Western Siberia: impacts of climatic warming on permafrost melting. *Intern. Journ. of Environ. Studies,* 66. 23-31.

Kling, G.W., Kipphut, G.W., Miller, M.C. (1991). Arctic lakes and streams as gas conduits to the atmopshere: implications for tundra carbon budgets. *Science* 251, 298-301.

Kravtsova, V.I., Bystrova, A.G. (2009). Study of thermokarst lake dynamics in Russia. *Geoinformatics*, 1, 44-51.

Kritzberg, E.S., Cole, J.J., Pace, M.L., Granéli, W., Bade, D.L., (2004), Autochtonous versus allochtonous carbon sources of bacteria: results from whole-lake 13C addition experiments. *Limnol. Oceanogr.* 49(2), 588-596.

Lehner, B., Doll, P. (2004). Development and validation of a global database of lakes, reservoirs and wetlands. *J. Hydrol.* 22(1). 296. (doi:10.1016 /j.jhydrol.2004.03.028).

Liss, O.L., Berezina, N.A. (1981). Mires of Western Siberia. Moscow, Moscow State University Press, 204 (in Russian).

Luoto, M., Heikkinen, R.K., Carter, T.R. (2004). Loss of palsa mires in Europe and biological consequences. *Environmental Concervation*. 31 (1), 30-37.

Malyasova, E.S., Novikov, S.M., Usova, L.I. (1991). Dynamics of peat accumulation and the process of mound bogs formation in Western Siberia // *Botanical Journal*. 76 (9), 1227-1237.

Matthews, J.A., Dahl, S.O., Berrisford, M.S., Nesje, A. (1997). Cyclic development and thermokarstic degradation in the mid-alpine zone at Leirpullan, Dovrefjell, Southern Norway. *Permafrost and Periglacial Processes*. 8: 107-122.

Melnikov, E.S., Veisman, L.I., Kritsuk, L.N. (1974). Landscape indicators of geocryological state of West siberian north and interpretive features. Moscow, Nedra. 132. (in Russian).

Mires of Western Siberia: their structure and hydrological regime. (1976). Ed. K. Ivanov and S. Novikov. Leningrad. *Hydrometeoizdat,* 448 (in Russian).

Mudrov, Yu.V., (2007). Permafrost conditions in cryolithozone of plains and mountains. Basic concepts and definitions. Moscow. Scientific World. 316. (in Russian)

Muldiyarov, E.Y. (1987). Frozen peatbog on the Ket-Chulym interfluve. *Glaciers and climate.* Tomsk, 84-85. (in Russian).

Novikov, S.M., Usova, L.I. (1983). Genesis of mound bogs of Western Siberia. *Questions of hydrology of bogs.* Leningrad, Hydrometeoizdat. 11-16 (in Russian).

Novikov, S.M., Usova, L.I., Malyasova, E.S. (1999). Age and dynamics of bogs of Western Siberia. *Bogs and boggy woods in view of tasks of sustainable wildlife management: Materials of meeting.* Moscow, 72-76 (in Russian).

Piavchenko, N.I. (1971). To studying of paleogeography of the West Siberian North in Holocene. *Holocene Palinology*. Novosibirsk, 139-157 (in Russian).

Polishchuk, Y.M., Tokareva, O.S. (2006). The monitoring of changes in the state of permafrost under warming using satellite images. *Bulletin of the Yugra State University*, 3. 87-90.

Popov, A.I. (1945). Origin of mound bogs of Taz-Yenisei interfluves. Abstracts of research works for 1944. Geology-geography Branch of RAS, Moscow-Leningrad (in Russian).

Romankevich, E.A., Vetrov, A.A. (2001) Cycle of carbon in the Russian Arctic Seas. M.: Nauka. 302. (in Russian).

Romanovsky, V., Armstrong, R., Hinzman, L.D., Oberman, N., Shiklomanov, A. (2007). Land. Arctic Report Card. [electronic resource]. – Mode access: http://www.arctic.noaa.gov/reportcard/land.html.

Samsonov, R., Lesnykh, V., Polishchuk, Yu., Bryksina, N.(2007). The climate change impact on thermokarst in West siberian territory and geological risks in gas industry // Proc. of 14th annual conference of TIEMS (June 5-8, 2007, Split, Croatia). – Split: TIEMS. - pp. 212 – 216.

Salmi, M. (1970). Investigations on palsas in Finnish Lapland. Ecology of the Subarctic regions. Proseedings of Helsinki Symposium 1966/ UNESCO, Paris, 143-153.

Schumilova, L.V. (1962). Botanical geography of Siberia. Tomsk: *Bulletin of Tomsk State University*, 1962. 439 (in Russian).

Seppala, M. (1988). Palsas and related forms. In: Advances in Periglacial Geomorphology, ed. M.J. Clark,. Chichester, UK: John Wiley and Sons. 247-278.

Shirokova, L.S., Pokrovsky, O.S., Kirpotin, S.N., Dupre, B. (2009) Heterotrophic bacterio-plankton in thawed lakes of the northen part of Western Siberia controls the CO_2 flux to the atmosphere. *International Journal of Environmental Studies*, 66:4,433-445.

Shirokova L.S., Kirpotin, S.N., Desmukh, C., Pokrovsky, O.S., Audry, S., Viers, J. (2013) Biogeochemistry of organic carbon, CO2, CH4, and trace elements in thermokarst water bodies in discontinuous permafrost zones of Western Siberia. *Biogeochemistry*, 113, 573-593.

Smith, L.C., Sheng, Y., McDonald, G.M., Hinzman, L.D. (2005). Disappearing Arctic Lakes, *Science*, 308, 1429.

Sobek, S., Algesten, G., Bergstrom, A.K., Jansson, M., Tranvik, L.J., (2003), The cathcment and climate regulation of pCO2 in boreal lakes. *Global Change Biology* 9, 630-641.

Tranvik, L., (1988), Availability of dissolved organic carbon for planktonic bacteria in oligotrophic lakes of differing humic content. *Microbiol. Ecol.* 16, 311-322.

Tyrtikov, A.P. (1979). Dynamics of a vegetative cover and development of frozen forms of relief. Moscow, 116 (in Russian).

Vasiliev, S.V. (2007). Forest and mire landscapes of Western Siberia. Tomsk, NTL, 276 (in Russian).

Walter, K.M., Zimov, S.A., Chanton, J.P., Verbyla, D., Chapin III, F.S. (2006). Methane bubbling from Siberian Thaw Lakes as a positive feedback to climate warming. *Nature* 443, 71-75.

Walter, K. M., Smith, L. C., F. Chapin S. (2007). Methane bubbling from northern lakes: present and future contributions to the global methane budget. Phil. Trans. R. Soc. A. 365, 1657- 1676 doi:10.1098 /rsta.2007.2036. Published online 18 May 2007.

Zemtsov, A.A. (1976). Geomorphology of West-Siberia plain (northern and central parts). Tomsk. Izd-voTomskogo Universiteta: 344 (in Russian).

In: Permafrost
Editor: Oleg S. Pokrovsky

ISBN: 978-1-62948-830-1

Chapter 6

MODELING AND PREDICTION OF DYNAMICS OF THERMOKARST LAKE FIELDS USING SATELLITE IMAGES

Vladimir Polishchuk*[1,2] and Yury Polishchuk[1,3]

[1]Tomsk State University
[2]Institute of Monitoring of Climatic and Ecological Systems, SB RAS
[3]Ugra Research Institute of Information Technologies

ABSTRACT

An analysis of experimental data on the statistical properties of the fields of thermokarst lakes in the area of permafrost in Western Siberia was carried out using the remote measurement of lake areas. On the basis of experimentally determined statistical properties of the thermokarst lake fields the geo-simulation model of the dynamics of these fields is proposed. The relationship of climatic and geocryological parameters changes was studied and multiple regression equation to take account of temperature and precipitation changes in the model was determined. Geo-simulation software package which allows the prediction of the dynamics of the thermokarst lake fields taking into account identified thermokarst and climate changes was developed. Application of the model will be shown in the examples of forecasting the dynamics of thermokarst lake fields until the year 2030 with the linear trend in mean annual temperature of air, given by the Intergovernmental Panel on Climate Change (IPCC). As a result of computer experiments with the model, it is shown that the lake areas in the West-Siberian permafrost in average will be reduced if the mean annual air temperature continues to increase in the coming decades.

Keywords: Geo-simulation modeling, relationship of climatic and geocryological changes, geo-simulation software, thermokarst dynamics prediction

INTRODUCTION

Intensification of geocryological processes in the permafrost caused by climate changes will be accompanied by intensified impact on infrastructure and other buildings in arctic regions initiating an increase in geoecological risks and resulting in large economic damage. Development of measures intended to lower the damages requires valid predictions concerning changes in permafrost state taking into account ongoing global warming. In these conditions getting expected assessments of changes in permafrost state under the impact of climate changes is an extremely important problem, for whose solution one should develop techniques and tools of mathematical modeling based on using experimental data concerning changes in permafrost state. The West Siberian wetland being hard to reach, collecting data is impossible without using the remote sensing of the earth surface.

As shown in the previous section, thermokarst lakes, which can be decoded easily in satellite images, are the most suitable indicators of geocryogenic surface changes in northern permafrost landscapes. Sets of lakes of different size form characteristic of these landscapes thermokarst lake fields that display noticeable changes in both space and time. Their coastal frontier shape and water surface area change, new lakes appear, some lakes disappear, turning into khasyreys (hollows of drained lakes). To develop techniques and tools for modeling and predicting thermokarst lake field dynamics it is necessary to find out statistical regularities of temporal-spatial changes of these fields' properties basing on the analysis of experimental data obtained with use of remote study of lake field dynamics. The research focuses on developing a mathematical model and designing means to predict thermokarst lake field dynamics in the permafrost.

1. Analysis of Experimental Data on Statistical Properties of Thermokarst Lake Fields

Experiments were carried out in three zones of permafrost in Western Siberia shown in Figure 1. The study used 106 cloud-free Landsat's satellite images selected from the USGS Global Visualization (http://glovis.usgs.gov). Data on areas of thermokarst lakes measured in satellite images obtained on 29 sections of the test sites, the location of which is shown in Figure 1.

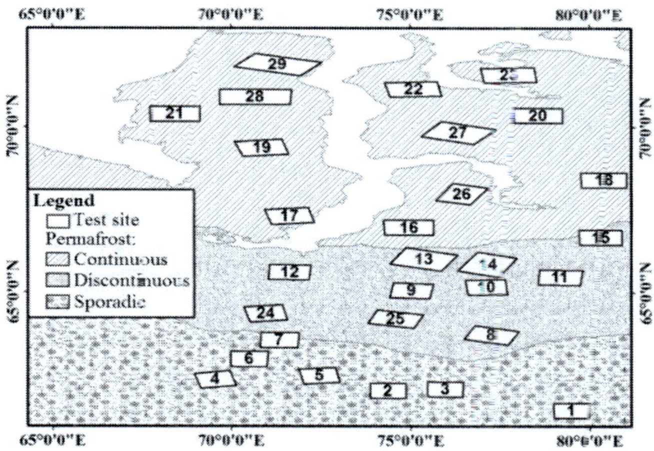

Figure 1. Geocryological map of West-Siberian lowland. The locations of test sites were chosen to evenly represent different permafrost types.

1.1. Experimental Regularities for Distribution of Characteristics of Thermokarst Lakes Location

To state the regularities for distribution of random coordinates of lakes satellite images were used. Figure 2 displays a fragment of decoded image from Landsat-5 (13.07.2007) of TS-9. After processing the image we got histograms of distribution of latitude and longitude values of location of lakes' centers shown in Figure 3, where specified:

K_i – relative number of lakes, occurring in every i -th histogram interval determined by the formula:

$$K_i = \frac{n_i}{N},$$

where – the number of lakes in every histogram interval;

i – interval number;

N – the whole quantity of lakes in the test site under study relevant to the given year.

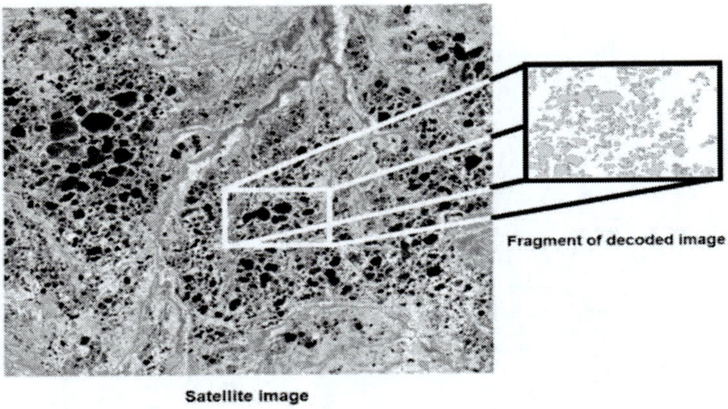

Figure 2. Sample of satellite image from Landsat-5 (13.07.2007) with fragment of a decoded image.

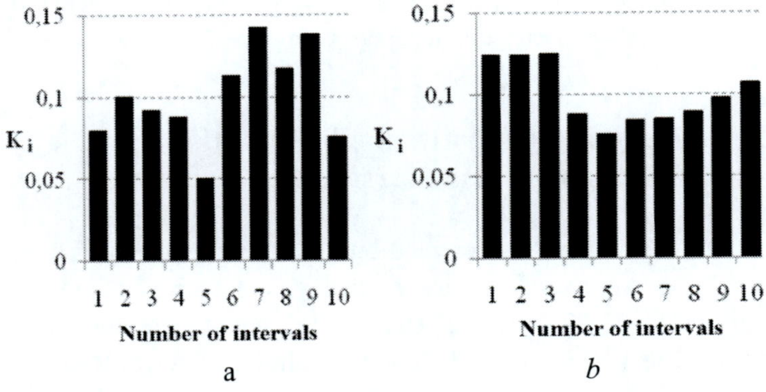

Figure 3. Histogram of distribution of latitude (*a*) and longitude values (*b*) of the lakes' location in TS-9.

The analysis of these histograms displayed that experimental regularities of distribution of coordinates of lakes' centers over the plane correspond to the law of uniform density according to criterion χ^2 with a probability of 95 % (Wentzel, 2002; Kremer, 2003).

1.2. Experimental Laws of Thermokarst Lakes' Distribution in Accordance with Their Areas

Histograms of distribution of lakes' number in accordance with their areas were built for all the test sites. For example in Figure 4 there are histograms for 3 test sites located in different permafrost zones.

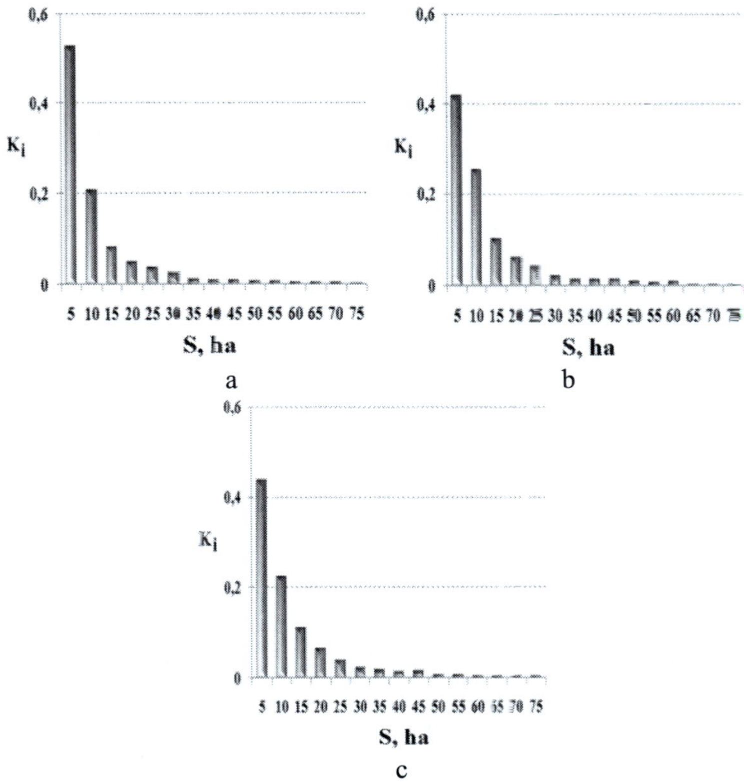

Figure 4. Histograms of distribution of lakes in accordance with their areas *a* – TS-1 (sporadic permafrost); *b* – TS-8 (discontinuous permafrost); c – TS-19 (continuous permafrost).

As comparison of the diagrams shows they have a common exponential character of the experimental law of distribution, which makes it possible (for thermokarst lake fields to be modeled easily) to choose one-parameter demonstrative law to describe thermokarst lakes' distribution in accordance with areas by the following formula:

$$y = \lambda \times e^{-\lambda s}, \tag{1}$$

where is a parameter of distribution law.

The value of parameter can be determined with the help of experimental data. According to Wentzel (2002), mathematical expectation of random quantity following the distribution law is determined in the form:

$$M(s) = \frac{1}{\lambda}.$$

Using as the estimate of mathematical expectation the value of average lakes' area estimated in accordance with experimental data by the following formula:

$$\bar{\mathrm{S}} = \frac{1}{\mathrm{n}} \sum_{i=1}^{n} S_i \,,\ i = \overline{1,n},$$

we will find that parameter can be determined with the help of experimental data according to the formula:

$$\lambda = 1/\overline{S} \tag{2}$$

Testing the correspondence of demonstrative law of lake area distribution to experimental histograms showed that in all TS this law corresponds to experimental data in accordance with criterion χ^2 with average probability of 90 %. Consequently, the above stated demonstrative law of lake distribution according to areas does not contradict the experimental data. The results of analyzing experimental distribution of lakes according to their areas displayed that λ in all the TS in a continuous permafrost varies in the range of 0.037 – 0.071, and in a discontinuous permafrost – in the range of 0.034 – 0.086 with average values 0.054 and 0.060 respectively.

1.3. Analysis of Statistical Relationship between Changes in Geographical Coordinates of Lakes' Centers and Areas

From the point of view of modeling thermokarst lakes it is interesting to investigate the statistical relationship between changes in thermokarst lakes' areas and coordinates of their centers. Table 1 contains correlation matrices determining the degree of statistical relationship between changes in geographical coordinates of the lakes and their areas. Here we use the following notations:

s – lakes' area;
x – geographical longitude;
y – geographical latitude.

Table 1 contains calculated values of correlation factors in TS-1, TS-8 and TS-19 located in different permafrost zones. Correlation factors between longitude and latitude changes are, in different permafrost zones basing on the data collected in TS-1, TS-8 and TS-19, 0.02, –0.06 and –0.08 respectively. Such low values of correlation coefficients between the changes of coordinates lead to conclusion about statistical independence of longitude and latitude changes. Correlation factors between values of areas and coordinates are within the interval –0.09 and +0.04, which also gives the grounds for a conclusion about the statistical independence of changes in lakes' areas and their centers' location coordinates in different permafrost zones.

Table 1. Correlation matrices of relationship between changes in thermokarst lakes' areas (*s*) and centers' coordinates (*x*,*y*) in different permafrost zones

Sporadic Discontinuous Continuous

random quantity	s	x	y	random quantity	s	x	y	random quantity	s	x	y
s	1	0.04	–0.07	s	1	–0.06	–0.09	s	1	–0.04	–0.03
x	0.04	1	0.02	x	-0.06	1	–0.06	x	–0.04	1	–0.08
y	–0.07	0.02	1	y	-0.09	-0.06	1	y	–0.03	–0.08	1

Consequently, an important feature of thermokarst lake fields is statistical independence of changes in random quantities of center coordinates and lakes' areas.

1.4. Remote Studying of the Shape of Thermokarst Lakes' Coastal Frontiers

The research was done in test sites located in different permafrost zones in West Siberia using satellite images from Landsat-4 (07.07.1988; TS-5); Landsat-5 (20.09.1989; TS-9); Landsat-4 (12.07.1988; TS-20). In each test site 10 thermokarst lakes of different size were chosen randomly. The lakes' areas were measured with the help of geoinformation system ArcGis 9.3. The results are presented in detail in Polishchuk V.Yu. and Polishchuk Yu. M. (2012).

For example, Figure 2 displays a fragment of a satellite image from Landsat-5 (20.09.1989) showing a field with thermokarst lakes of different size. As one can see in Figure 2, thermokarst lakes usually have a round shape, and we are going to show the possibility of presenting their frontiers' shape as a circle. For example, in Figure 5 there is a satellite image of a real thermokarst lake's coastal frontier.

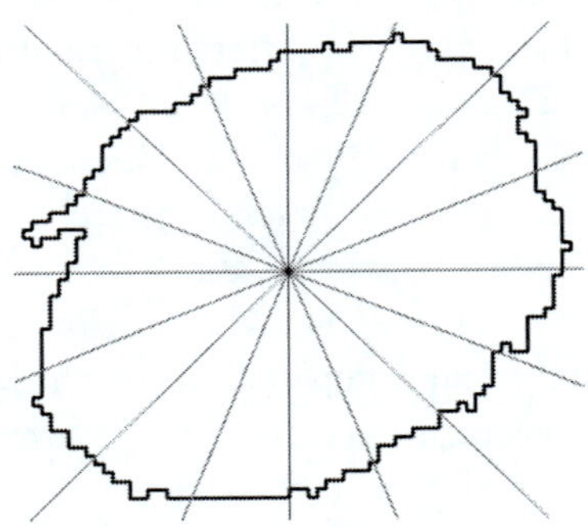

Figure 5. A satellite image of a thermokarst lake with a ray-path plot.

In our research (Polishchuk V.Yu. and Polishchuk Yu. M., 2012) we present the results of measuring segments of radial lines for 10 lakes chosen randomly in TS-9. Analysis of the data showed that root-mean-square error is 0.048. Consequently, a relative error of replacing lakes' coastal frontier line with a circle does not exceed 5 %. Similar processing the results of measurement segments of radial lines for 10 lakes in other TS showed that the relative error of replacing lakes' coastal frontier line with a circle is 0.049 in TS-5 and 0.052 in TS-20 respectively.

Thus, research conducted in test sites in sporadic, discontinuous and continuous permafrost showed that the error in estimating lakes' areas while replacing their real coastal frontier line with a circle is comparatively small (about 5 %). It may serve as a reason to choose a circle as a model for a lake in simulation modeling thermokarst lake fields.

2. Development of a Geo-Simulation Model of Thermokarst Lake Fields

In the mathematical modeling of thermokarst lakes are used analytical and simulation models. According to published works, mathematical modeling of thermokarst processes exploits analytical models. The analysis (Matt, 2011) of the results of the research in thermokarst processes which was done using various analytical models revealed that these models are quite efficient in studying processes in a single thermokarst lake but unsuitable for modeling thermokarst lake fields. In this respect, the methods of mathematical landscape morphology developed by A.Victorov (Victorov, 2009; Victorov, 2006) are of great importance here as they are designed to use analytical models for territory dynamics modeling.

A.Victorov (Victorov, 2006) researched development of thermokarst-erosion plains using a mathematical model of their morphological structure. A characteristic feature of their structure is a combination of generation expansion processes of thermokarst lakes, on the one hand, and khasyreys appearance, on the other hand. It is noted, that a test site within a long period of development reaches the state close to static, i.e., the state characterized by dynamic balance (Victorov, 2006). All the above means that active centers of thermokarst processes are constantly appearing and disappearing on permafrost territory. Though their shapes and sizes are changing, basic numerical statistical characteristics of location density of active thermokarst centres and their size remain steady. Therefore, methods of mathematical morphology of landscape allow to predict changes in the territory state over significant development period (Victorov, 2009), accounting hundreds of years, using analytical knowledge. Methods and models of mathematical morphology of landscape can be used in geoecological risk estimations for thermokarst process development. However, the methods of mathematical morphology of landscape are not designed to study spatial-temporal changes in thermokarst lake fields under the conditions of current global warming for the last 2-3 decades.

Consequently, judging by the review of the works devoted to mathematical modeling of thermokarst processes, at present, there are no models allowing to study spatial-temporal changes of random thermokarst lake fields. This fact triggered development of mathematical models of random thermokarst lake fields within simulation modeling.

Below are basic properties of geo-simulation model of thermokarst lake fields. As it has been discussed in the previous section modeling of thermokarst processes has not been fully developed yet, and a model of thermokarst lake field has not even been designed. According to simulation modeling methodology (Low, 2004; Shannon, 1978), a simulation model of the object under study must consider basic properties of this object. The present section describes results of an experimental research of the properties of thermokarst lake fields using remote-sensing methods.

Considering the information stated above, the following fundamental principles determining substantial properties of a model of spatial-temporal structure of thermokarst lake field can be formulated:

1. Lake coastline shapes can be represented by a circle equation with centers coordinates x_i, and area (i - lake serial number).
2. Spatial changes in the position of centers of circles and their areas are statistically independent.
3. Random distribution of circle centers coordinates is governed by a uniform law.
4. Random distribution of number of circles over their areas conforms with the exponential law of distribution as in (2.1) with as a parameter.
5. Time changes in statistical properties of aggregate of random circles and their dependence on climate changes are determined by dependency of parameter on time and climatic characteristics in the following equation:

$$\lambda = f(T, \text{P}, t),$$

where T is temperature, is level of precipitation and t is time.

A model of thermokarst lake field is a aggregate of random circles (Figure 6), whose statistical characteristics correspond to the above principles (1-5). Geometrical interpretation of the model of thermokarst lake fields is presented in Figure 6. Using number triple (x, y, s), representing the value of the center of the circle coordinates and value of its area, coordinates of the points determining borders of each circle are calculated in the following way:

$$x_k = R * \cos\gamma + x, \qquad (3)$$

$$y_k = R * \sin\gamma + y, \qquad (4)$$

where and – coordinates of the center of the circle;

x_k and – coordinates of k – th point on the circle;

γ – value of axial angle x and radius, directed from the center of the circle into k –th point on the circle

R – radius of the circle, computed using the following formula $R = \sqrt{s/\pi}$;

where – random value, distributed by exponential law and determining the area of the circle.

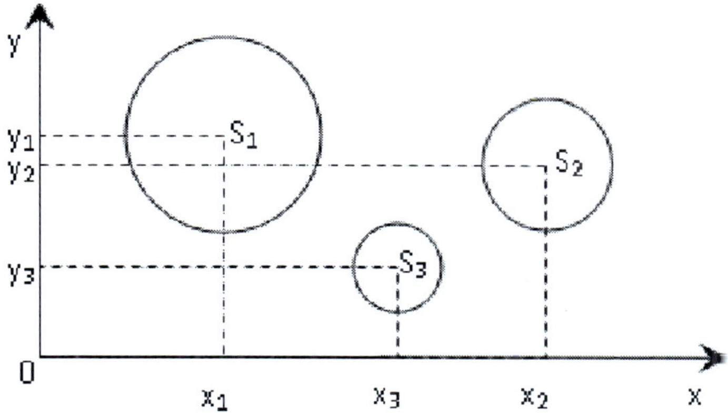

Figure 6. Geometrical representation of thermokarst lake field model fragment as an aggregate of random circles.

Consequently, the basic elements of model description are characteristics covering lake shapes, parameters of their random location on surface and random distribution of lakes over their size (areas). Analysis of experimental data on number change of thermokarst lakes, presented in the previous section, showed that the relative number of re-emerged lakes does not exceed 9% for the last three decades on the test site, while a relative number of disappeared lakes over the same period is a per cent share. Therefore, lake appearance and disappearance processes were not reflected in the developed geo-simulation model of thermokarst lakes.

3. Correlation of Climatic and Geocryological Changes in Permafrost Zones and Its Accounting in the Model

3.1. Study of Climate Changes in West-Siberian Permafrost Zone

Whereas it is necessary to consider climate change influence on properties of thermokarst lake fields, the study of correlation of climatic and thermokarst changes in West Siberia perennial permafrost zone in modeling presents a paramount interest. To study this correlation a simultaneous analysis of data on area dynamics of thermokarst lakes and time sequences of climatic indices provided by meteorological stations located on the geocryological changes test sites. To study interrelation of climatic and geocryological changes location overlap (or at least close location) of test sites and meteorological stations locations is necessary. According to the analysis of test sites and meteorological stations locations on the territory under study, locations of meteorological stations and test sites, as a rule, don't match. It is known that meteorological stations are usually based on river and sea coasts or in inhabited locality on hard-to-reach territories of perennial permafrost zone and thus are quite distant from test sites. Such location divergence of test sites and meteorological stations can cause considerable inaccuracy in the results of analysis of correlation of climatic and geocryological changes on the tested territory. As such, to analyze correlation of area change of thermokarst lakes and climatic indices (average annual temperature and precipitation level) an alternative approach was taken to obtain data on air temperature and precipitation. The approach is based on meteorological data reanalysis which allowed to estimate the value of climatic indices in test sites. Reanalysis (repeated meteorological analysis) (Meteorological reanalysis, 2011) – is a means to obtain meteorological information in set points of the tested territory. Reanalysis is based on "acquisition" of historical observational data from the net of meteorological stations, covering continuous period and using a unified scheme of consistent data "acquisition" over the whole period of analysis. The main advantage of such meteorological data is uniform cover of the territory.

It should be noted that various systems of reanalysis are used by scientists nowadays. In the present work *Reanalysis ECMWF ERA-40 and APHRODITE JMA* systems were used to define climatic characteristics. A brief description of reanalysis data use procedure in the present research will be presented further. Average monthly values of air temperature and annual precipitation in test sites from 1970 till 2010 were obtained with the help of reanalysis. As reanalysis data are represented in a set of graphic files in the well-known *.png format, to extract information necessary for climatic study, a template map locating test sites was developed. The created template map was overlaid on the map of temperature and level of precipitation fields, obtained by reanalysis, and with the help of graphic editor MSPaint a new image representing an overlap of test site location maps and temperature (or precipitation level) field was received. Such integrated maps helped to determine average monthly values of air temperatures and annual precipitation in the centre of each test site. The next stage was to calculate average value of air temperature for each test site monthly and annually. As a result, tables of time series of annual average value of air temperature and annual precipitation for each test site were obtained. Study of statistical correlation of temperature changes in different permafrost zones ("friendliness" of temperature changes) is seen to be of some interest for the research. Correlation coefficients of average annual values of air temperature for each permafrost zone were calculated. The results of correlation coefficients calculation between temperature changes in different permafrost zones are: sporadic and discontinuous – 0.94; sporadic and continuous – 0.78; continuous and discontinuous – 0.92.

3.2. Studying Interrelation between Geocryological and Climate Changes in the Permafrost

As it was shown in the previous section features of thermokarst lake fields change under the conditions of contemporary climate warming. The most sensible indicator of thermokarst processes influenced by temperature rise is thermokarst lakes' area, which is illustrated with the results of remote study in West Siberia (Dneprovskaya et al., 2009; Smith et al., 2005). The analysis of interrelation between thermokarst and climate changes is given in (Bryksina et al., 2009; Polishchuk V. Yu. et al., 2009). This section is aimed at describing the results of analysis of interconnection between geocryological changes in thermokarst lake fields and climate features (air temperature and precipitation level) in the permafrost in West Siberia.

Further one can see the results of comparing time series of average lake area and average annual temperature in 29 test sites. According to the results of measuring lakes' areas with the help of satellite images average lake areas were determined in each test site. The analysis of time series of total areas of thermokarst lakes allowed stating trends in changing average area of the lakes. For example in Figure 7 there is a diagram of time dependence of average area of thermokarst lakes in TS-10, displaying on average decrease in time in average lakes' area with coefficient of a linear trend $\alpha = -0{,}11$ha/year. A straight line in Figure 7 presents the trend line.

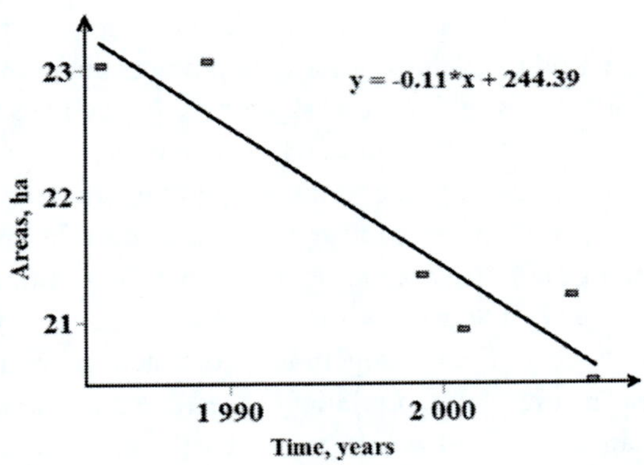

Figure 7. Changes in the average area of thermokarst lakes in time.

To study the interrelation between the changes in thermokarst lakes' areas and changes in air temperature and precipitation level we compared (Polishchuk V. Yu., Polishchuk Yu. M., 2013) coefficients of a linear trend of time changes in average values (for full territory) of thermokarst lakes' areas and climate characteristics.

Comparing time dependence trends showed that the increase in average annual temperature in the last decades coincides with decrease in lakes' areas. The uncovered average decrease in lakes' areas can be explained by the impact of climate warming on permafrost for the last 3 – 4 decades. The previous section of the book contains explanation for the trend of decrease in lakes' areas and description of possible mechanisms of their decrease under the global warming influence. A similar trend for decrease in thermokarst lakes' areas in other permafrost territories is marked in (Kravtsova and Bystrova, 2009).

However, it is the analysis of the data obtained aimed at developing a model of thermokarst lake fields suitable for predicting that is of most interest. To do this it is necessary to study climate dependence of parameter λ, which determines the kind of the law for thermokarst lakes' distribution in accordance with their areas discussed in section 1.2. Table 2 contains the data of changes in parameter λ and climate features. It should be pointed out that the data exist only for the years when cloudless images were taken, which made it possible to calculate the value of parameter λ.

The diagrams illustrating dependence of the value of parameter λ on average annual air temperature and precipitation level presented in Figure 8 display dependence of parameter λ on climate indicators under study.

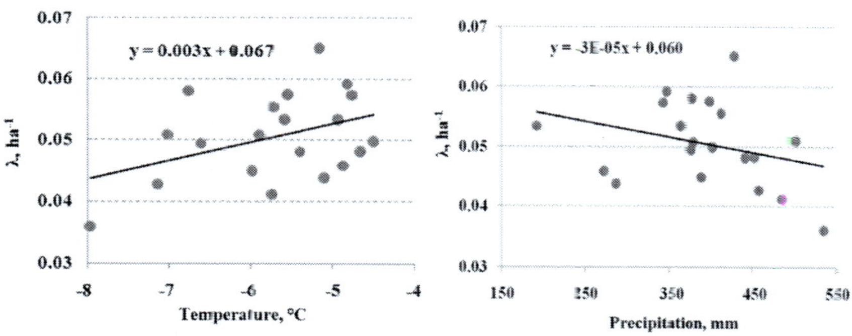

Figure 8. Dependence of the values of parameter on annual rainfall and average annual air temperature.

To develop a model of thermokarst lake fields' dynamics it is necessary to define this dependence in explicit form, which was the reason for doing multidimensional regression analysis (Kremer, 2003) of time series of the values of parameter and climate features in the West Siberian territory under study given in Table 2.

The results of multidimensional regression analysis of the data from Table 2 can be presented in the form of equation of multiple regression in the form:

$$\lambda = c_0 + c_1 * x_1 + c_2 * x_2 + c_3 * x_3, \tag{5}$$

where – average annual air temperature, – precipitation level, – time, – coefficients of regression equation, $i = 0,...,3$.

In the result of the regression analysis of time series of the values of parameter λ and climate features given in Table 2 the following values of regression equation coefficients were obtained (5):

c_0 = -0.585 ha^{-1}; c_1 = 0.00062 ha^{-1}/°C; c_2 = 0.000014 ha^{-1}/mm; c_3 = 0.00032 ha^{-1}/year.

Table 2. Time series of parameter and climate features

Year	Parameter λ, ha^{-1}	Average annual air temperature,° C	Annual precipitation, mm
1953	0.036	–7.97	535.49
1973	0.041	–5.74	486.81
1977	0.049	–6.60	376.76
1981	0.050	–4.50	403.10
1983	0.048	–4.66	453.28
1984	0.045	–5.98	389.63
1985	0.058	–6.76	378.60
1987	0.051	–7.02	379.82
1988	0.048	–5.40	442.87
1989	0.058	–5.55	399.53
1992	0.043	–7.14	458.96
1999	0.051	–5.90	503.04
2000	0.055	–5.72	413.22
2001	0.053	–5.59	364.22
2002	0.065	–5.16	429.13
2003	0.044	–5.10	287.90
2006	0.053	–4.93	192.37
2007	0.046	–4.87	273.44
2008	0.059	–4.81	347.45
2009	0.057	–4.75	344.03

The stated regression dependence of parameter λ on time and climate changes is a basis for developing algorithms for modeling random thermokarst lake fields, which is discussed in the following section.

4. GEO-SIMULATION MODELING AND PREDICTION OF THERMOKARST LAKE FIELDS

4.1. Developing an Algorithm for Modeling a Spatiotemporal Structure of Thermokarst Lake Fields

In a general case mutual density of probabilities of random coordinates of centers and areas of circles imitating lakes in a mathematical model of random thermokarst lake fields can be presented in the form:

$$f(x, y, s), \quad (6)$$

where and – coordinates of circle center in a model;

s – area of a circle imitating a lake.

Consequently, the totality of circles in the model of lake fields will be presented as a totality of groups of three random values (x, y, s). To develop an algorithm for modeling thermokarst lake fields it is necessary to take into consideration statistic connections between changes in lakes' coordinates and their areas. The analysis of correlation matrix done in subsection 1.3 showed that random spatial changes in areas and coordinates of lakes' centers in the same way as coordinates of lakes' centers between each other are statistically independent. This is the reason why, when modeling thermokarst lake fields, mutual density of probabilities (3) can be presented in the form:

$$f(x, y, s) = f(x) \times f(y) \times f(s), \quad (7)$$

where and – densities of the probability of even distribution; and value of model parameter are determined by (1) and (2) respectively.

Taking into account equation (5) random-number sequence determining characteristics of location of circles' centers (x and y) is generated using antenna of pseudorandom numbers distributed in accordance with the law of even distribution. And to form circles of random size whose areas are distributed according to the law conforming to equation (1) it is necessary to generate random-number sequences distributed in accordance with demonstrative law conforming to equation (Buslenko, 1968):

$$s_i = -\frac{1}{\lambda} \ln z_j, \tag{8}$$

where – numbers with even distribution in the interval (0,1), $j = 1,...,m$.

Consequently, together with using software generators for even distribution of pseudorandom numbers software realization of imitation model of thermokarst lake fields includes creating generator for pseudorandom number sequences distributed in accordance with demonstrative law. Then algorithms for modeling spatiotemporal structure of random thermokarst lake fields are considered.

We should consider geo-simulation model of spatial structure of a thermokarst lake field that is a totality of circles and reflects the state of a thermokarst lake field at the moment of time t, whose fragment is shown in Figure 6. Further this model is called a particular model for a thermokarst lake field. To model the dynamics of thermokarst lake fields we should consider a general model of spatiotemporal structure of a thermokarst lake field in the form:

$$M_{ПВ} = \{M_{П}(t_1),...,M_{П}(t_j),...,M_{П}(t_n)\}, \; j = 1,...,n, \tag{9}$$

which is a totality (time sequence) of particular geo-simulation models of a thermokarst lake field $M_{П}(t_j)$, $j = 1,...,n$, where each particular model relates to a particular moment of time.

Figure 9 gives visual presentation of the general model for spatiotemporal structure of thermokarst lake fields in the form of geoinformation system (GIS) layers that relate to given time moments $t_1, t_2, \ldots, \mathrm{t_n} \in \overline{(\mathrm{t_1}, t_n)}$.

While modeling spatiotemporal structure of thermokarst lake fields it is important to take into consideration both time dependence and climate features (temperature, precipitation level). In view of the above the dependence of parameter λ on time and climate features is determined by the equation of multiple regression in the form (2.5). This is the reason why equation (2.5) was used to develop an algorithm for numerical modeling dynamics of thermokarst lake fields.

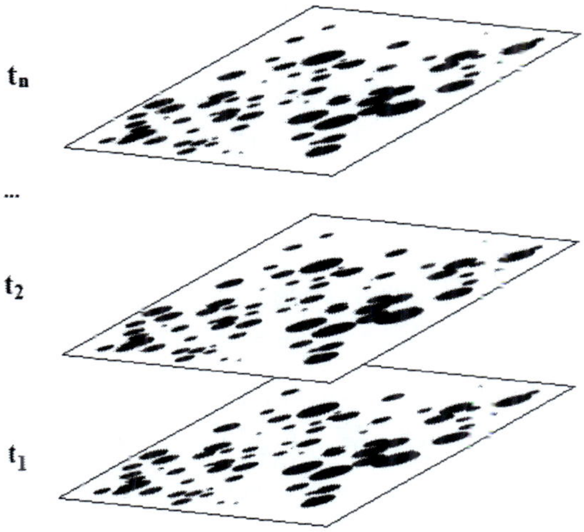

Figure 9. Visual presentation of a general model of dynamics of thermokarst lake fields. Le end: – time (year), $i = \overline{1,n}$.

The developed algorithm for modeling dynamics of thermokarst lake fields can be presented as the following:

step 1 – the year of modeling is specified t_j, $j = 1,...,m$;

step 2 – the areas (S_{MA}) of the model area (*MA*) under study are specified;

step 3 – lake density (δ_{MA}) in *MA* is specified;

step 4 – the number of circles within *MA* is determined in accordance with formula: $N_{MA} = S_{MA} \times \delta_{MA}$;

step 5 – the center of *MA* location in the map is specified;

step 6 – parameter is determined in accordance with formula (5) for given values of temperature and time (year of modeling);

step 7 – pseudorandom number is generated, distributed in accordance with uniform law;

step 8 – using the number obtained at the previous step pseudorandom number is calculated according to formula (8) to characterize the value of circle area;

step 9 – two pseudorandom numbers are generated, distributed in accordance with uniform law, determining the coordinates for circle center location on the screen;

step 10 – using the values of a number triple (x, y, s) obtained at previous steps 8 and 9, in accordance with equations (3) and (4) a circle is formed on the screen;

step 11 – If the number of circles obtained is less than N_{MA}, determined at step 4, the algorithm repeats beginning with step 7, otherwise it is completed.

The given algorithm allows forming a particular model of spatial structure for a given time moment $M_{\Pi}(t_j)$, where $j = 1,...,m$. To make a general model of dynamics of a thermokarst lake field by means of forming time sequence of particular models $M_{\Pi}(t_j)$ for a given set of moments t_j (*i* = *1,…,m*) the algorithm repeats for the number of times (m) needed.

4.2. Software Package for Geo-Simulation Modeling Thermokarst Lake Fields

A structural scheme of the developed software package for geo-simulation modeling thermokarst lake fields is presented in Figure 10. It contains the following components (Polishchuk V. Yu., 2013):

1) the subsystem of modeling thermokarst lake fields;
2) the subsystem of displaying results;
3) the database (DB).

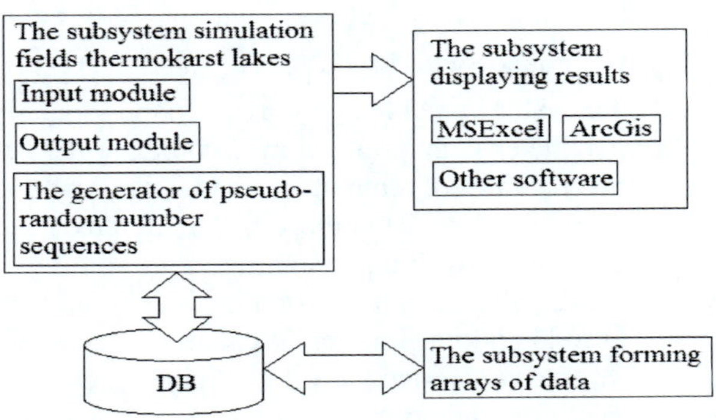

Figure 10. A general scheme of software package for geo-simulation modeling thermokarst lake fields.

Input module is intended to provide the generator for pseudorandom number sequences with the values of model parameters.

The generator of pseudorandom number sequences is intended to form pseudorandom number sequences while realizing algorithms of modeling thermokarst lake fields.

Output module is intended to convert modeling results into one of the following formats:

- Microsoft Excel (*.xls);
- Vector format (*.shp);
- Raster format (*.jpeg).

With the help of the subsystem of displaying modeling results output data can be displayed either in a digital map by means of geoinformation system (for example, ArcGIS), or as an electronic table and graph in MSExcel.

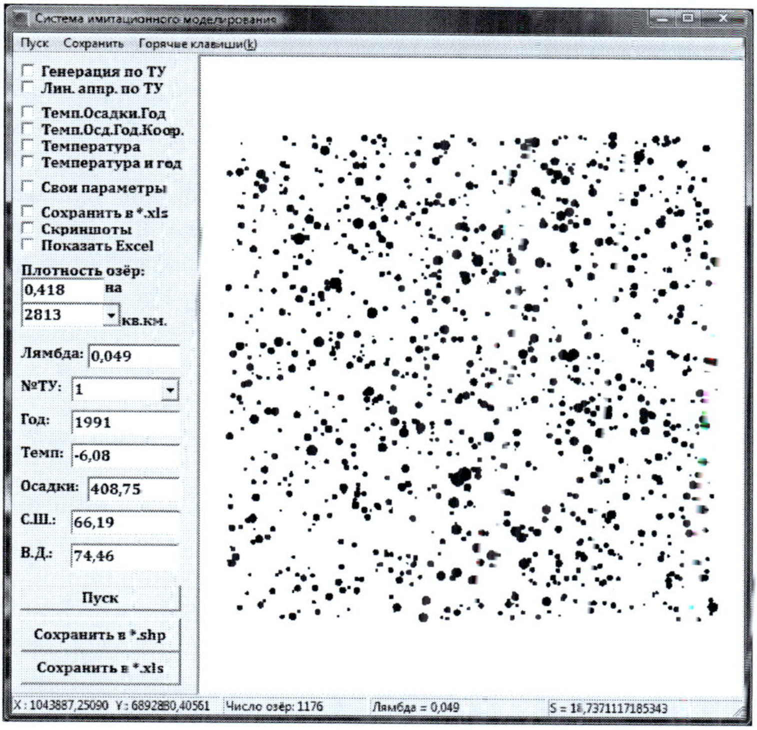

Figure 11. Form view of software package for geo-simulation modeling.

Database (DB) is storage for spatial and attributive information about the properties of thermokarst lakes under study collected in the result of processing satellite images as well as the data concerning changes in climate parameters in the territory under study.

It should be said that software realization of the algorithm for geo-simulation modeling was done in the environment Delphi7. The generator for pseudorandom numbers distributed in accordance with uniform law used in the research is the one installed in the development environment Delphi7. Figure 11 presents an example of displaying a thermokarst lake field model in developed software package for modeling.

The software package for geo-simulation modeling thermokarst lake fields is registered by Russian Federal Intellectual Property Service (certificate № 2011614293 dated May, 31, 2011) and available from the authors upon request (Polishchuk V. Yu., 2011).

4.3. Studying Model Validity and Modeling Accuracy

The research is based on carrying out computer-based experiments. Model validity was studied by comparing average lake areas according to experimental and modeling data. Two data arrays containing 29 values corresponding to the number of TS were compared. The first data array contains averaged for the period under study average values of thermokarst lake areas for each TS, determined in accordance with satellite images. The second one – averaged for the period under study average values of model lake areas calculated with the help of the developed model. As far as the second data array is concerned it should be mentioned that in accordance with empirical data about averaged for the period under study average values of lake areas the values of parameter λ are determined, which were used in modeling.

Comparison of two selected arrays was done in accordance with Student test. The formula used in calculation (Comparing group…, 2011) to compare two small samples i.e., samples less than is:

$$t_{\beta} = \frac{\left|\overline{X} - \overline{Y}\right| * \sqrt{n * (n-1)}}{\sqrt{\sum\left[(x_i - y_i) - \left(\overline{X} - \overline{Y}\right)\right]^2}}$$

with the number of degrees of freedom $k = 2*(n-1)$,

where – value of Student's test, used to determine confidence probability of groups' difference;

x_i and – averaged for the period under study average values of lake areas in i-th TS according to experimental and modeling data respectively for i-th TS;

$\overline{X}$ and – arithmetic means for averaged for the period under study average values of lake areas according to experimental and modeling data;

n – size of experimental and modeling data samples.

During the research it was stated that with number of degrees of freedom = 56 value of Student's test = 0.62. According to (Comparing group…, 2011), if = 0.62 < = 2,003 with = 56, then for confidence probability 95 % we can make the conclusion about unreliability of difference between the samples. Consequently, the developed model can be regarded appropriate to experimental data.

Accuracy of modeling dynamics of thermokarst lake fields was studied in the form of simulation experiment on the model in accordance with scenario 1.

Scenario 1. *Modeling dynamics of a thermokarst lake field basing on experimental data about average values of lakes' areas averaged for all the test sites in the permafrost under study.*

Values of parameter in this case are calculated according to multiple regression formula (5) using the data of average annual temperature and precipitation level determined for each TS by reanalysis. Then a model field is formed in accordance with the algorithm described above. The formed model thermokarst lake field is displayed on-screen (Figure 11), then the array of values of modeled lakes' areas is saved as a file for it to be analyzed.

Figure 12 presents graphs of time dependence of experimental and calculated values for parameter λ. A solid line in Figure 12 displays time dependence of calculated values for parameter λ, and a dashed line represents time dependence of average values of experimental quantities λ. Sections of vertical dashed lines in Figure 12 show the swing (intervals between maximal and minimal values) of experimental average values of λ. In Figure 12 one can see that the graph of modeled time dependence keeps within the boundaries of intervals of changes in experimental data, which indicates good correspondence between modeled and experimental data of parameter λ.

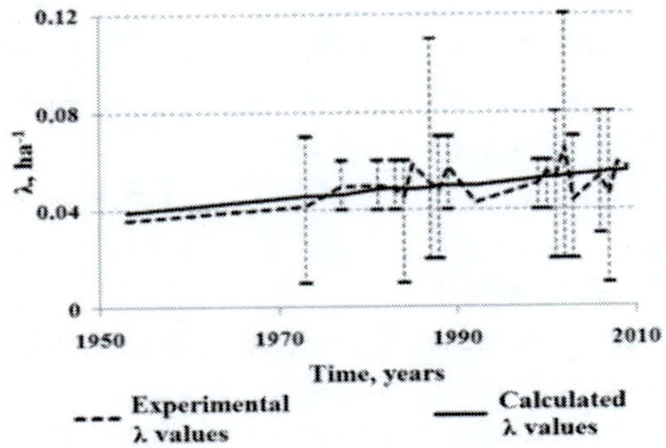

Figure 12. Time dependence of experimental and modeled values of parameter λ.

As it was mentioned above, calculated values of were used to generate model lake fields used to determine average values of model lake areas. The graph of time dependence of model area values is shown with a solid line in Figure 13. A dashed line shows time dependence of average experimental values of lakes' areas. Vertical dashed sections indicate intervals between maximal and minimal values of experimental average values of lakes' areas. Figure 13 shows that the graph for average values of model areas in general keeps within the intervals, which may serve to indicate modeling results' reliability.

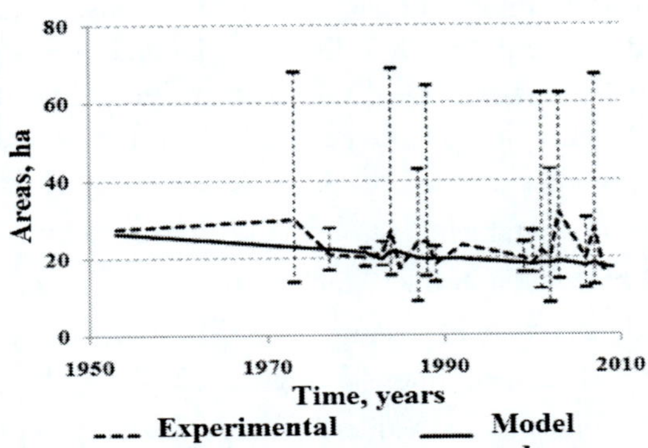

Figure 13. Time dependences of average values of experimental and modeled average values of lakes' areas.

To estimate the accuracy of modeling average thermokarst lake areas in the permafrost we shall determine root-mean-square deviation with formula (Wentzel, 2002):

$$\sigma = \sqrt{\frac{\sum_{i=1}^{n} (\frac{x_i - y_i}{x_i})^2}{n-1}},$$

where and – average values for experimental and modeled lake areas respectively;

n – number of values, $i = \overline{1, n}$.

Estimation showed that error of determination of average values of lakes' areas on base of modeling with use of experimental data in accordance with scenario 1 is 17 %, which may be well regarded as a suitable result of modeling thermokarst lake fields for predicting thermokarst lake fields dynamics.

5. Predicting Spatiotemporal Changes in the Fields of Thermokarst Lakes

The thermokarst lakes changes prediction is based on simulation experiments with the model according to a particular scenarios. The development of such a scenario involves prediction of changes in the average annual air temperature and precipitation level on the permafrost area in the predicted period of time from now until 2030. Analysis of the reputed predictions of climate changes in the coming decades (Strategic forecast…, 2005; Pachauri et al., 2007; Klimenko et al., 2007; Khrustalev et al., 2008) showed that temperature in the forecast period of 2010-2030 on average matches the extrapolation of a linear trend of a time course of temperature during the previous observation period (1970-2000). So in the future, when determining the expected values of temperature in modeling the dynamics of thermokarst lakes until 2030, we will use extrapolation of the linear trend of the time course of average annual temperature obtained for the permafrost area under study in Western Siberia based on the reanalysis data.

Prediction of the thermokarst lake fields' dynamics was conducted using the above described complex with the help of computer experiments with the model. To assess the quality of prediction a prediction interval is determined (Confidence interval, 2011), which constitutes a range of points, in which the predictand values are expected to be found when conducting additional observations (with a given probability). We should note that, unlike prediction interval, confidence interval (Confidence interval, 2011), provides information about the expected (mean) value of the dependent variable, and determines a range of possible values for the expected value of the predicted variable, which will be located around the "true" mean value of the dependent variable with a given level of confidence probability. Prediction intervals in this paper were calculated using the software package STATISTICA for the 95% probability.

Scenario 2. *Prediction of the thermokarst lakes fields dynamics on the basis of extrapolation of trends of temperature and precipitation changes using ERA-40 and APHRODITE JMA reanalysis data for the period of up to the year of 2030.* This scenario uses the temperature data obtained with the help of ERA-40 reanalysis system and the data on changes in precipitation calculated with the Japanese system of reanalysis APHRODITE JMA. The prediction is for the period from 2010 to 2030. The values of the average annual air temperatures were obtained by extrapolating the linear trend of the average annual temperature time course, represented by a dashed line on the graph in Figure 14.

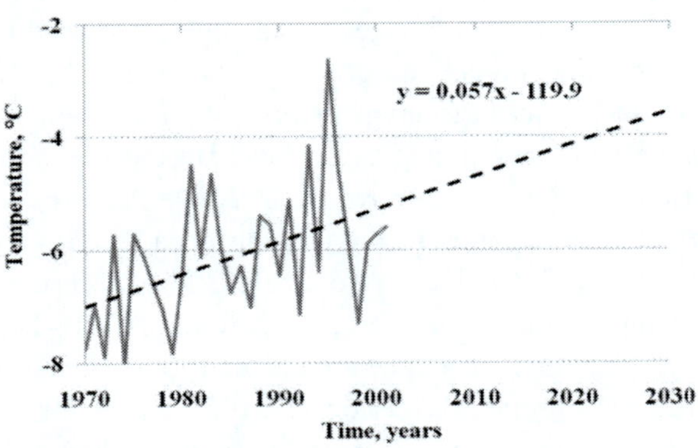

Figure 14. Extrapolation of the linear trend of the average annual air temperature time course in the permafrost area of Western Siberia.

The time dependence of the expected values of model parameter is shown in Figure 15, where circles show the expected values according to the regression equation, and the dashed lines represent prediction intervals, calculated for the probability of 95%. A simulation experiment was conducted according to scenario 2, aimed at obtaining thermokarst lakes model fields in the predicted period of 2010-2030. In thermokarst lakes fields modeling parameter expected values for the predicted period were used (Figure 15). The model lakes average areas were determined by modeling using simulation software package for thermokarst lake fields geo-simulation modeling.

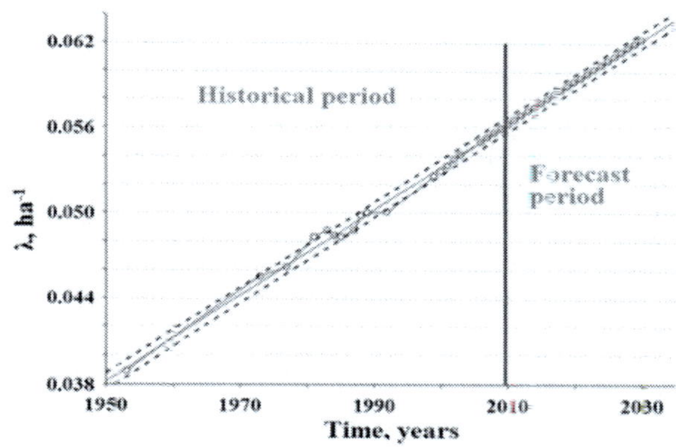

Figure 15. The time dependence of the expected values of parameter for the period of 2010-2030.

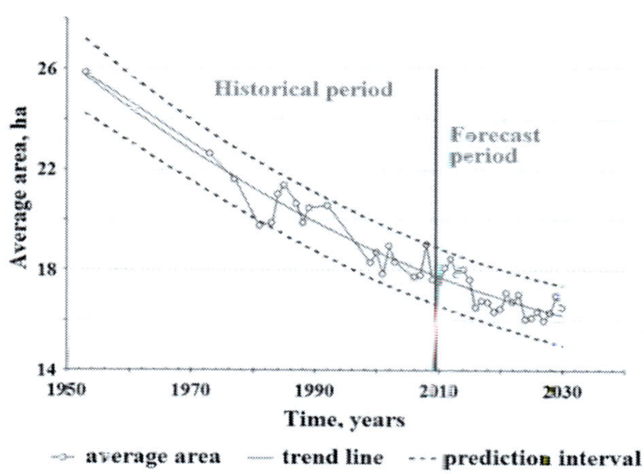

Figure 16. Predicted time dependence of the lake area average values.

The predicted time course of average values of thermokarst lakes model fields gained as a result of modeling is shown in Figure 16, where circles show the model values of the lakes areas, and dashed lines represent prediction interval calculated with a probability of 95%. The solid line in the middle of the prediction interval represents trends of the average lakes areas time dependence calculated using the STATISTICA application package.

ACKNOWLEDGMENTS

This research was supported by Russian Federal Programme “Kadry” (contract No 14.740.11.0409), ESA CAT-1 Project ID:5762 “Cryogenic processes of monitoring in Russian permafrost territories using radar and optical space images”, RFBR Project No. 12-05-90833-mol_rf_nr and the Megagrant of Russian Federal Programme 14.B25.31.0001 (BIO-GEO-CLIM).

REFERENCES

Bryksina, N. A., Polishchuk, V. Yu., Polishchuk,Yu. M. (2009). Study of relationship between climate change and thermokarst processes in the areas of continuous and discontinuous permafrost in Western Siberia / *Bulletin of the Yugra State University*. 3, 3-12. (in Russian)..

Buslenko, N. P. (1968). Simulation of complex systems. Moscow. *Nauka*. 356. (in Russian).

Comparing groups using Student's test. (2011) [electronic resource]. - Mode of access: http://www.kgafk.ru/kgufk/ html/uchmetrologia7.html (in Russian).

Confidence intervals and prediction intervals. (2011) [electronic resource]. - Mode of access: http://www.statsoft.ru/home / portal / glosary / glossarytwo / C \ ConfidenceIntervalsPredictionInterval.htm (in Russian).

Dneprovskaya, V. P., Bryksina, N. A., Polishchuk, Yu.M. (2009). Study of changes of thermokarst lakes in the area of the discontinuous permafrost in Western Siberia on the base of satellite images. *Study of Earth from Space*. 4, 88-96. (in Russian).

Klimenko,V.V., Khrustalev, L.N., Mikushina, O.V., Emeliyanova, L.V., Ershov, E.D., Parmuzin S.Yu., Tereshin A.G. (2007). Climate change and dynamics of the permafrost in north-western Russia within the next 300 years / *Cryosphere of Earth*. 11(3), 3-13. (in Russian).

Khrustalev, L.N., Klimenko, V.V., Emeliyanova, L.V., Ershov, E.D., Parmuzin, S.Yu., Mikushina, O.V., Tereshin, A.G. (2008). Dynamics of permafrost temperature in southern regions of cryolithozone under different scenarios of climate change // *Cryosphere of Earth*. 11(1), 3-11. (in Russian).

Kravtsova, V.I., Bystrova, A.G. (2009). Study of thermokarst lake dynamics in Russia. *Geoinformatics*, 1, 44-51. (in Russian).

Kremer, N. S. (2003). Theory of probability and mathematical statistics: a textbook for high schools Moscow. UNITY-DANA. 573. (in Russian).

Low, A.M., Kelton, W.D. (2004). Simulation: Classic Computer Science. St. Petersburg-Kiev. Publishing Group BHV. 847. (in Russian).

Matt, A. (2010). Modeling thermokarst lakes dynamics and carbon flux. [electronic resource]. - Mode of access: http://www.docstoc.com/docs/35351317/Methods_of_thermokarst_lakes_modeling

Meteorological reanalysis. (2011) [electronic resource]. - Mode of access: http://en.wikipedia.org/wiki/ (in Russian).

Pachauri, R.K., Reisinger, A. IPCC. (2007). Climate Change: Synthesis Report. Contribution of Working Groups I, II and III to the Fourth Assessment Report of the Intergovernmental Panel on Climate Change. IPCC, Geneva, Switzerland. 104.

Polishchuk, V. Yu. (2011). Program simulation of space-time structure of the fields of thermokarst lakes. Registered Rospatent - Certificate number 2011614293. (in Russian).

Polishchuk,V. Yu. (2013). Software system of dynamics simulation of thermokarst lake fields in permafrost zones // *Reports of TUSUR*. 1, 125-128. (in Russian).

Polishchuk, V. Y., Bryksina, N. A., Dneprovskaya, V. P., Polishchuk, Y.M. (2009). Analysis of the relationship of climatic and geocryologic changes in the permafrost of West Siberia. Proceedings of the 8th Siberian conference on climatic and ecological monitoring (Tomsk, 8-10 October, 2009). Tomsk. Agraf-Press. 34-36. (in Russian).

Polishchuk, V. Yu, Polishchuk, Yu. M. (2012). Remote studies of variability of the shape of coastal boundaries of thermokarst lakes in the permafrost of West Siberia // *Study of Earth from space*. 1, 61-64 (in Russian).

Polishchuk, V. Yu., Polishchuk, Yu.M. (2013). Geo-simulation modeling fields of thermokarst lakes in zone of permafrost. Khanty-Mansiysk // *Bulletin of Yugra State University*. 129. (in Russian).

Shannon, R. (1978). Simulation systems - the art and science. Moscow. *Academic Press.* 411. (in Russian).

Smith, L.C., Sheng, Y., MacDonald, G.M., Hinzman, L.D. (2005). Disappearing Arctic Lakes // *Science*. 308(3), 14.

Strategic prediction of climate changes of the Russian Federation for the period up to 2010-2015 years and their impact on the sectors of the Russian economy.(2005). Moscow. *Roshydromet.* 30. (in Russian).

Victorov, A. S. (2006). The main problems of mathematical morphology. Moscow. *Nauka.* 252. (in Russian).

Viktorov, A. S. (2009). Mathematical Models of surface morphology in the geoecologic task decision. / Sergeevskie reading. Geo-ecological modeling in solving problems: Materials of annual session of the Scientific Council on Environmental Geoscience, Engineering Geology and Hydrogeology (Moscow, 23-24 March, 2009). GEOS, *Issue* 11. 154-159. (in Russian).

Wentzel, E. S. (2002). Probability theory. Moscow. *Vyschaya School.* 575. (in Russian).

In: Permafrost
Editor: Oleg S. Pokrovsky

ISBN: 978-1-62948-830-1

Chapter 7

IMPACT OF PERMAFROST THAW ON THE BIOGEOCHEMISTRY OF THE THERMOKARST LAKES IN WESTERN SIBERIA: CURRENT STATUS AND POSSIBLE FUTURE CHANGES

O. S. Pokrovsky*, L. S. Shirokova, R. M. Manasypov, S. N. Kirpotin, S. P. Kulizhsky and S. N. Vorobiev
GET UMR 5563 CNRS, University of Toulouse (France)
and BIO-GEO-CLIM Laboratory, Tomsk State University (Russia)

ABSTRACT

Western Siberia's thermokarst (thaw) lakes extended over a territory spanning over a million km²; they are highly dynamic hydrochemical systems that receive chemical elements from the surrounding peat soil and exchange greenhouse gases with the atmosphere, delivering dissolved carbon and metals to adjacent hydrological systems. Because more permafrost will continue to thaw due to climate warming processes, which are heavily intensified in this region the directions and magnitude of these processes may be significantly modified, seriously affecting the biogeochemical fluxes both on land and in the coastal zone of the Arctic Ocean. In this chapter, we will analyze the biogeochemical and morphological features of thermokarst lakes along the sequence of their natural evolution: from frozen palsa depressions to large, km-size lakes and drained lakes. Based on a compilation of ~120 analyses of these water bodies, we report the average concentration of organic carbon, as well as the major and trace elements, in Western Siberia's thermokarst lakes; we will also predict the development of their chemical composition, exchange with the atmosphere and effect on the riverine fluxes that run from the land to the ocean under the climate warming scenario. The accelerating permafrost thaw and rising water temperatures in this region will likely shorten the life cycle of the thermokarst thaw water bodies, increase the fluxes of both CO_2 and CH_4 into the atmosphere, increase the concentration and delivery of dissolved organic carbon and related trace metals to the hydrological network and increase the potential bioavailability of micronutrients. The retroactive link between climate change and the thermokarst lakes' biogeochemistry may be more significant than is currently expected.

Keywords: Siberia, permafrost, thermokarst, lake, biogeochemistry, metal, carbon

INTRODUCTION

The western Siberia plain comprises more than 1 million km^2 of peat bog and lake landsape is most likely the terrestrial system most vulnerable to on-going climate change. The main reasons for this high vulnerability include the following: (*i*) discontinuously distributed permafrost with very low thickness relative to Central and Eastern Siberia (sometimes less than 50 meters); (*ii*) the dominance of thick (up to 3 m in the south and 1 m in the north) organic peat deposits that are far more reactive than mineral soils; (*iii*) high coverage of the territory by tundra with patches of forest along the river valleys; (*iv*) abundant thermokarst (thaw) lakes that often cover up to 80% of the watershed area; (*v*) dense infrastructure comprising roads, pipelines and settlements linked to expanding oil and gas production. Individually, each of these factors may produce great instability in western Siberian ecosystems relative to other permafrost zones due to global warming. When combined, these factors make this site a unique battlefield between atmospheric CO_2 drawdown processes (peat formation and vegetation production) and CO_2 release into the atmosphere due to soil and water organic matter degradation.

Thawing of the permafrost and releasing ancient organic carbon is one of the most significant environmental threats within the global climate change scenario (Schuur et al., 2008). Solid organic matter's degradation and subsequent conversion into gaseous CO_2 is facilitated by the liquid media. Numerous studies of permafrost degradation in subarctic Alaska (Jorgenson et al., 2011), Canada (Payette et al., 2004; Turetsky et al., 2002; Laurion et al., 2010) and Sweden (Christensen et al., 2004) have confirmed that the ecosystems are becoming wetter in the course of the permafrost disappearance.

Regardless of the mechanisms responsible for C_{org} → CO_2(gas) conversion, this process is enhanced in the surface waters, especially in the oxic environments. Numerous shallow surface depressions filled with thawed water and growing thermokarst (thaw) ponds and lakes in Western Siberia offer the most favorable natural environments for these conversion processes and therefore deserve thorough investigation.

The annual CO_2 emission fluxes from the thermokarst lakes in Western Siberia (45±27 mol C/m²/y, Ripo et al., 2007; 8.9±4.5mol C/m²/y, Shirokova et al., 2009) are similar to those from other thermokarst lakes and rivers in the NE European Russia (5-9 mol C/m²/y, Heikkinen et al., 2004). These values are on the same order of magnitude as those for aerobic respiration in the Arctic water column, its temperate lakes (0.6 – 10 mol C/m²/y, Ramlal et al., 1994 and references therein) and lakes in Alaska (12 mol C/m²/y, Kling et al., 1992). The CO_2 emission from lakes in Western Siberia are comparable with the heterotrophic respiration of thermokarst wetlands (16 mol C/m²/y, Wickland et al., 2006) but an order of magnitude higher than the amount of CO_2 emitted from the Canadian peatlands (0.73 mol C/m²/y, Turetsky et al., 2002; 0.4-1.4 mol C/m²/y, Vitt et al., 1990) and small boreal non-permafrost lakes (3.5 mol C/m²/y, Kortelainen et al., 2006). However, thermokarst lakes and ponds are important parts of the carbon budget at high latitudes because the typical CO_2 emissions from a lake's surface to the atmosphere (110±50 g C/m²/y, Shirokova et al., 2009 and references therein) are comparable to the typical values for the carbon sinks in peat plateaus (e.g., 106-110 g C/m²/y, Heikkinen et al., 2004). Because the newly formed lakes in the Western Siberia make up a high geographical proportion of the area (up to 48% of the surface area (Zimov et al., 1997; Hinkel et al., 2003; Riordan et al., 2006), reaching 60-80% in the Pur Taz and Nadum river basins (Zakharova et al., 2009), there is little doubt that the Western Siberian thermokarst lakes are likely to act as a major regulator for the CO_2 release from the frozen peat into the atmosphere.

Compared to large number of studies devoted to the biogeochemistry of aqueous and gaseous organic carbon forms in permafrost-affected soil and water bodies, relatively little is known regarding trace elements (TE). Recently, we have described the evolution of both the concentration and speciation of colloidal TE in the water column and in the sediments along the chronosequence of thermokarst lakes and ponds developing in the northern portion of western Siberia within continuous permafrost zone (Pokrovsky et al., 2011; Shirokova et al., 2013). However, to extend this knowledge to other boreal regions that are less affected by permafrost, studies of TE biogeochemistry in discontinuous permafrost zone under various landscape and geographical settings remain necessary. Today, it is critical to ascertain the extent that the TE's concentration in the thermokarst lakes is controlled by the stage of lake development, the landscape (forest, tundra and bog lakes of the watersheds compared to coastal lakes for example) and latitudinal position (climate and the permafrost thickness). Moreover, it is impossible to quantify the effect of permafrost thawing on the degree of chemical composition change of thermokarst water bodies and their potential role as greenhouse gas sources/sink under climate warming scenario, even as a first order approximation. Elucidating these issues should allow the large-scale modeling of the stocks and fluxes of the chemical elements within the aquatic reservoirs of the entire boreal permafrost-affected zone.

Study Sites, Sampling and Analyses

The study site is located in the central and northern portions of Western Siberia (63.5-70.5°N, near the towns of Nojabrsk and Novy Urengoy towns, as well as the Khanymey and Tazovsky settlements), within a primarily tundra landscape that has developed continuous and discontinuous permafrost over Neocene sand and clay deposits that are covered by a 1–2 m thick peat layer (Figure 1 A). In these regions, numerous thermokarst (thaw) lakes are observed that range in size from several hundred m^2 to km-sized lakes, as illustrated by the topographical map of a typical watershed divide area (the Khanumey test site used for this work) (Figure 1 B). An aerial view of the study site also confirmed the lakes' dominant land coverage (Figure 1 C).

The sampling and filtration, as well as the chemical analysis, techniques are very similar to those used during our previous studies (Pokrovsky and Schott, 2002; Pokrovsky et al., 2006; Vasyukova et al., 2010; Pokrovsky et al., 2010). An ultraclean sampling procedure was used throughout the manipulations in the field (Shirokova et al., 2010). The water samples were collected in a PVC boat from the central part of the lake or manually from a depth of 30-50 cm and immediately filtered through sterile, single-use Minisart® filter units (Sartorius, acetate cellulose filter) with 5 or 0.45 μm pore sizes. In addition to the conventional filtration, *in-situ* dialysis experiments were performed using 20-50 ml pre-cleaned dialysis bags placed directly into the lake. This dialysis procedure lasted between 72 and 96 hrs. The filtered and dialyzed samples were stored at 4–5°C in the dark before analysis. The concentrations of dissolved organic carbon (DOC), Cl, SO_4, cations and TEs, as well as alkalinity were measured using routine methods for analyzing boreal water samples in our laboratory (Vasyukova et al., 2010; Pokrovsky et al., 2010, 2011).

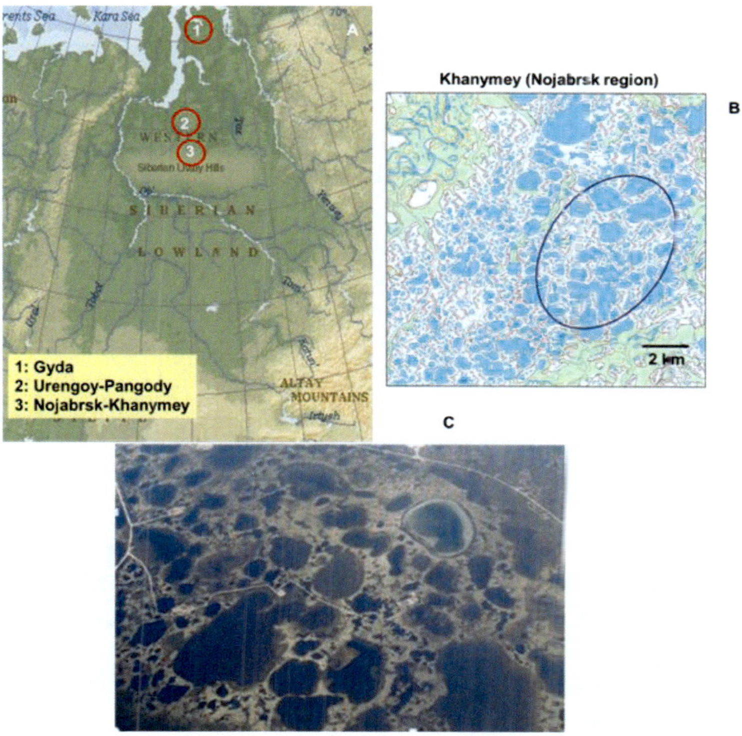

Figure 1. Schematic representation of study sites: a general map of western Siberia (A), the topographical map of the main study site (Khanymey, discontinuous permafrost zone, B) and an aerial photo of thermokarst lakes field in the southern part of continuous permafrost zone (C).

The dissolved inorganic carbon (DIC) concentrations were obtained from the alkalinity after a standard HCl titration procedure using an automatic Schott TitroLine alpha $TA10^{plus}$ titrator with an uncertainty of ±2% and a detection limit of 5×10^{-5} M. The DOC was measured using a Shimadzu TOC 6000 with an uncertainty of 5%. The trace elements were measured using inductively coupled plasma-mass spectrometry (ICP-MS, Agilent 7500 CE and Element XR) with indium and rhenium as the internal standards and a precision better than ±5 %.

The international geostandard SLRS-4 (Riverine Water References Material for Trace Metals certified by the National Research Council of Canada) was used to check the validity and reproducibility of each analysis. Good agreement was obtained between our replicated SLRS-4 measurements and the certified values (relative difference <10%).

RESULTS

Cyclic Development of Thaw Lakes on the Frozen Palsa Peat Bogs

The water bodies of the Western Siberian lowland follow specific life cycles, beginning from the appearance of m²-size soil depressions and permafrost subsidences and progressing to large, mature lakes that are subjected to drainage, as illustrated in schematic diagrams with photos corresponding to each stage in Figure 2.

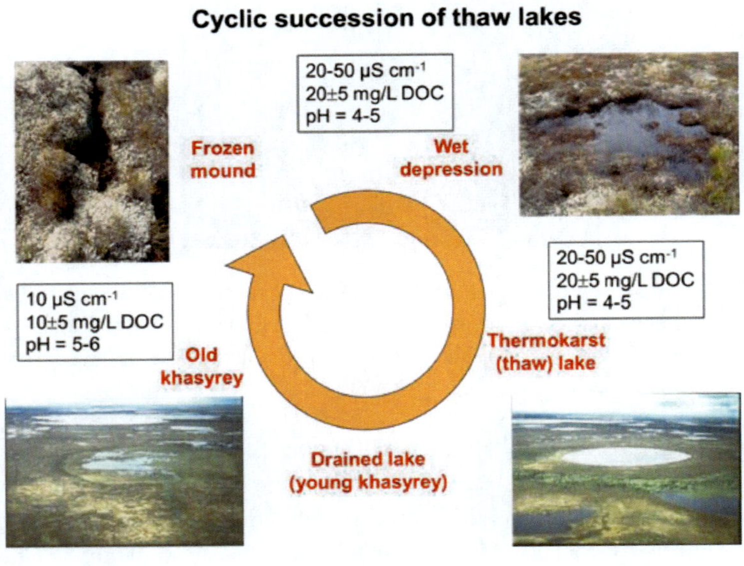

Figure 2. Scheme of cyclic succession of thermokarst lakes development in discontinuous permafrost zone of western Siberia.

In the dry bottom of large drained lake, a depression and permafrost subsidence may occur, initiating a new cycle of lake development. The chemical composition of thermokarst water bodies changes systematically along the timeline of lake development; the highest total dissolved solid (TDS) concentrations and dissolved organic carbon (DOC) occur at the beginning when the thaw ponds form within the permafrost subsidences and wet depressions (Figure 2). The largest mature lake with a km-scale diameter exhibits the lowest specific conductivity (close to 10 μS cm^{-1}, among the lowest known on the Earth surface for such a large and stable hydrological object), while the highest concentration of both DOC and metals are observed in the smaller meter-sized depressions.

3.2. Landscape Context-Based Types of Lakes and Their Chemical Composition

An important observation regarding the thermokarst lakes studied in this work is that all of them have ice (permafrost) under their beds below 30-50 cm (depending on the season) of porous organic-rich sediment. As such, the lakes are strongly isolated from the mineral substrate and deep groundwater discharge at the lakebed is minimal. The latter is confirmed by the extremely low total dissolved solid concentrations in km–sized lakes. The main source of DOM and inorganic solutes in the thermokarst lakes is the peat abraded at their borders; this abrasion may occur either by wave action or thawing/submersion actions on the portions on the border line, as illustrated by the ground photos in Figure 3 A and B, respectively.

In addition to the "classic" shallow thermokarst lakes most abundant in the Western Siberain arctic and subarctic, deep "primary" lakes of glacial or tectonic origin were also investigated. Figure 4 summarizes the hydrochemistry both large and small thermokarst lakes and presents the variations in the main physical and chemical parameters of two primary lakes located within the permafrost zone; these lakes were sampled during the summer baseflow season in 2010.

It can be seen from these figures that there is significant temperature stratification during summer season. The oxygen concentration gradually decreases from the surface to the bottom with only first 2 to 4meters being fully saturated with atmospheric O_2.

Figure 3. Photos of the pet abrasion at the thermokarst lake border in discontinuous permafrost zone (Nojabrsk region, 2010).

The gas regime of the two lake types is presented in Figure 5 and the metal concentrations relative to depth are provided in Figures 6 A-I.

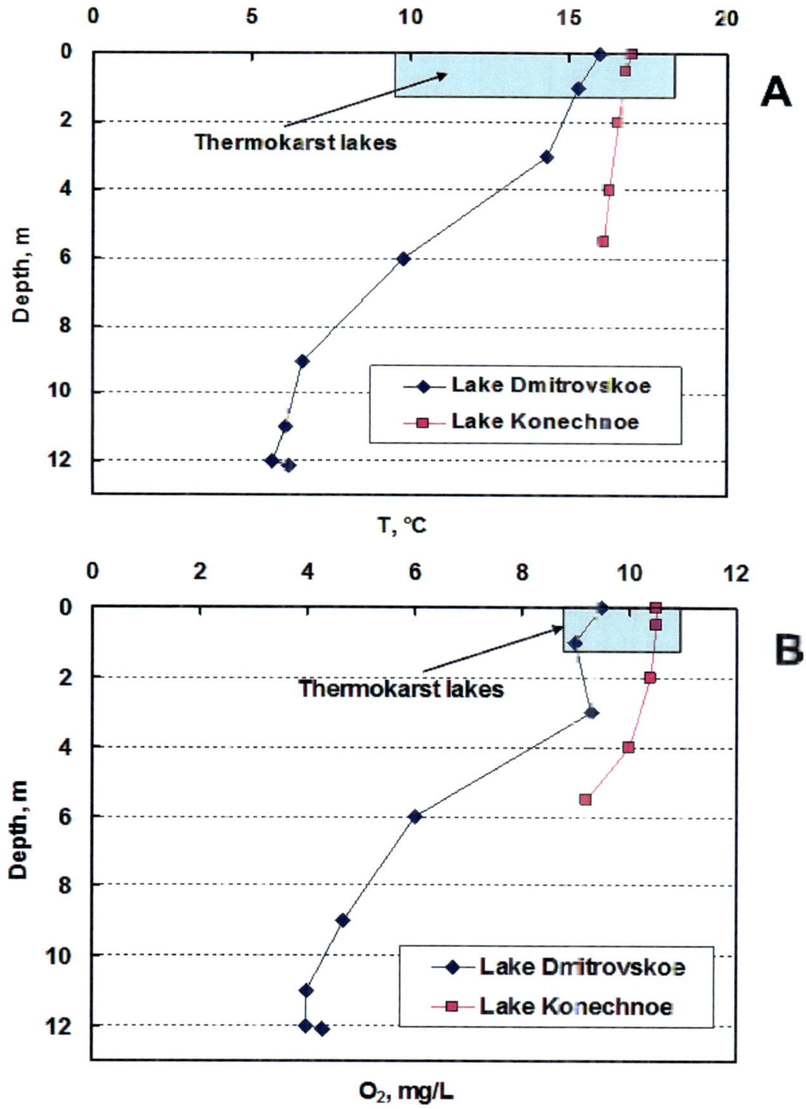

Figure 4. Temperature (A) and oxygen concentration (B) depth profile of usual shallow thermokarst lakes located in discontinuous permafrost zone (rectangular) and two deep "primary" lakes of sporadic permafrost zone shown by symbols with connecting lines.

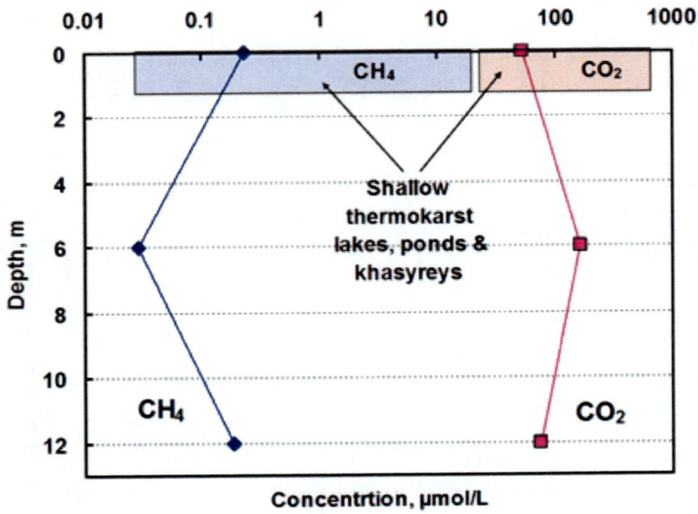

Figure 5. Methane and CO_2 concentration in shallow thermokarst lakes of western Siberia in discontinuous permafrost zone and depth profile of these gases in deep « primary » Dmitrovskoe lake (sporadic permafrost).

There is a significant difference between the specific electric conductivity of the lake water in the permafrost setting between the thermokarst and deep "primary" lakes; however, the latter exhibits one of the lowest specific conductivities known for isolated and stable water bodies on the Earth's surface. The concentration of methane and CO_2 in the deep lakes is similar to those in the shallow thermokarst lakes (Figure 5). The metal concentrations for most of the shallow thermokarst lakes are similar or comparable to those of the deep lakes. These low concentrations of total dissolved solid and lithogenic metals in the deep lakes located within the permafrost-dominated ground support the lack of detectable groundwater feeding the thermokarst lakes' chemical composition; the hypothesis suggested in earlier is valid for continuous/discontinuous permafrost terrain (Pokrovsky et al., 2011). The frozen state of the surrounding substrate (Neogene sands and peat deposits) is responsible for the low mineralization of surface waters in Western Siberia. The deep lakes located within the forest/forest tundra zone may also originate from large, thawed ice wedges/lenses because they exhibit extremely low concentrations of dissolved inorganic salts.

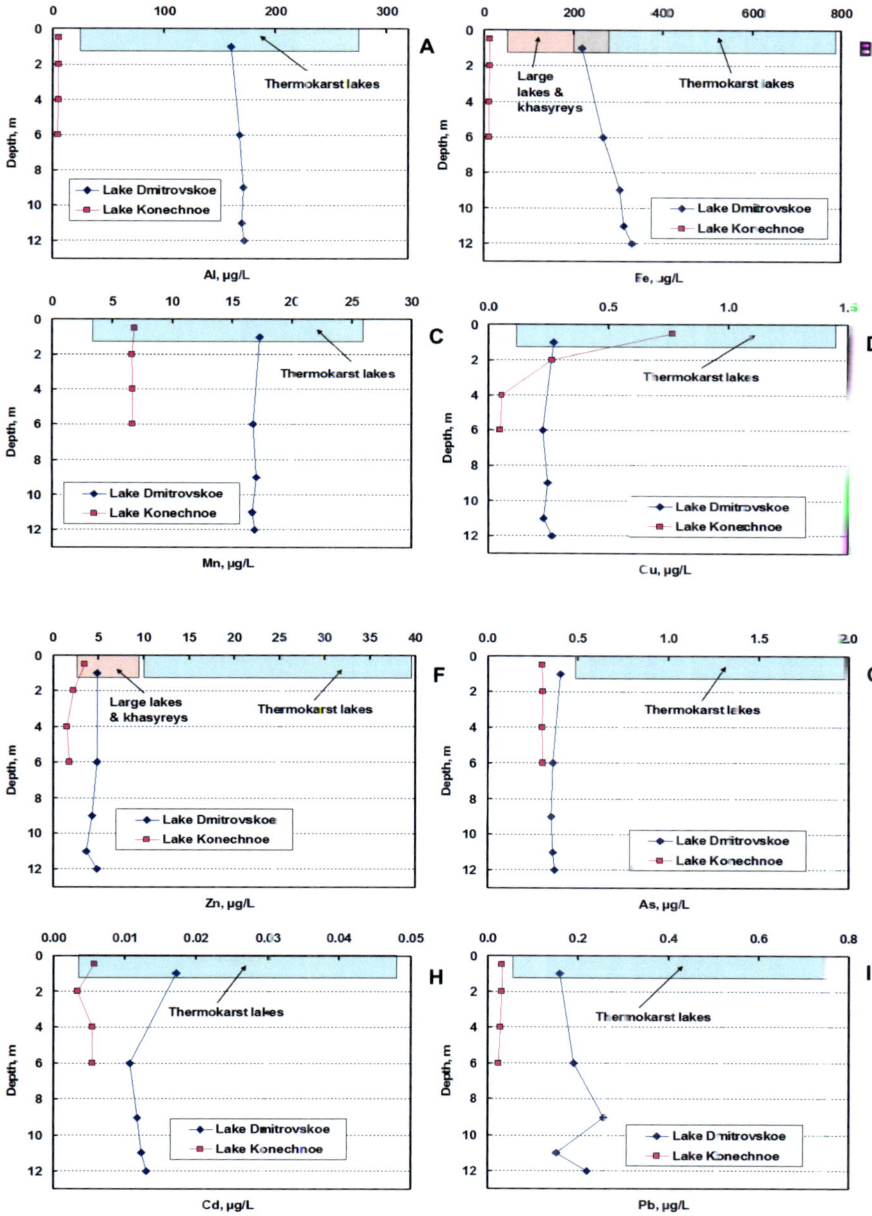

Figure 6. Trace metal concentration in shallow thermokarst lakes of western Siberia. See legend Fg. 4 for details.

Testing the Latitude (Climate and Permafrost) Gradient of Lake Chemical Composition

Given the large geographic coverage of the thermokarst lakes (10° latitude along a strongly pronounced gradient of permafrost thickness ranging from discontinuous and sporadic in the south of Nojabrsk district to continuous in the Gyda region), the effect of the climate or latitudinal position on lake water hydrochemistry can be determined. This representation may be partially biased because the Gyda lakes are located near the sea and are thus subjected to the influences of marine aerosols depositions. For this analysis we used the geometric mean of the element concentration of the lakes sampled during the summers of 2008-2012 for four size ranges of thermokarst water bodies: depressions smaller than 10 m diameter, thaw ponds less than 100 m diameter, small and medium lakes 100 to 500 m diameter (the dominant type of water bodies in the region) and mature lakes at the final stage of their development larger than 500 m in diameter.

We analyzed the major components of the lake water (DOC, Ca, Fe, Al), as well as the most environmentally important heavy metals (Zn, Mn, Cu, Ni and Pb).

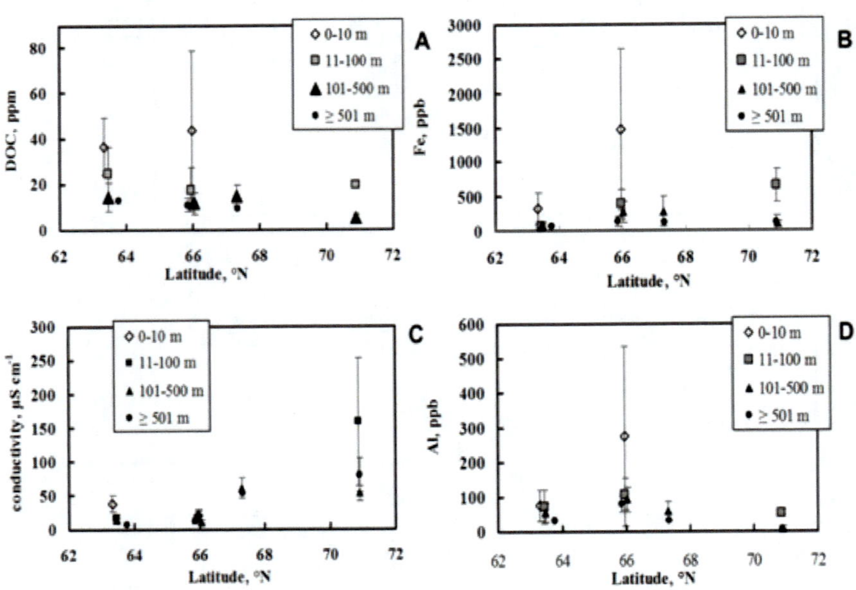

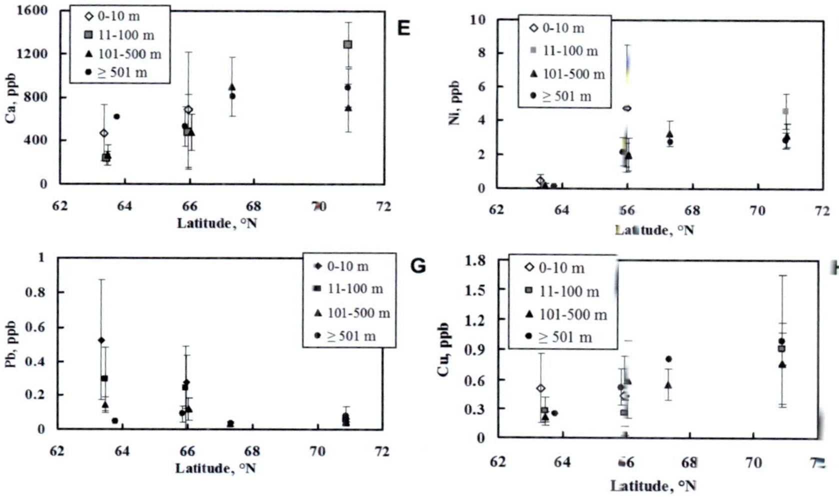

Figure 7. Latitude plot of DOC (A), Fe(B), specific conductivity (C) and Al (D) concentration for lakes of different size ranges, sampled in discontinuous / sporadic, discontinuous and continuous permafrost zones. Latitude plot of Ca (E), Ni (F), Pb (G) and Cu (H) concentration for lakes of different size ranges, sampled in discontinuous / sporadic, discontinuous and continuous permafrost zones.

There are 4 distinct patterns of elemental concentrations in the lake water relative to the lake's latitudinal position, as illustrated in Figures 7 A-I: (*i*) A decrease in the elemental concentrations from the south to the north (DOC, Pb), (*ii*) nonsystematic concentration evolution with the latitude (Fe and Al), (*iii*) a maximal concentrations at the continuous permafrost sites in the continental zone (Mn); (*iv*) an increase in the concentrations from south to north, notably in the coastal Gyda zone (specific conductivity, Ca, Zn, Cu, Ni, and Cr). A pronounced increase of Cu, Ni and Cr concentration in the most northern region of continuous permafrost zone (Gyda site) may be due to the proximity (a distance of ~ 300 km) from the largest arctic smelter, the Cu-Ni-Cr processing factories in Norilsk.

3.4. Impact of Extreme Climate Events on Thermokarst Lakes and Ponds Water Hydrochemistry

In summer 2012, a heat wave affected western Siberia, producing mean daily temperatures that were the highest measured since the beginning of the

20th century. Both the minimal and maximal temperatures of the summer in 2012 were 10-15 degrees higher than the previously studied summer (2010); the summer of 2010 exhibited normal conditions (Figure 8).

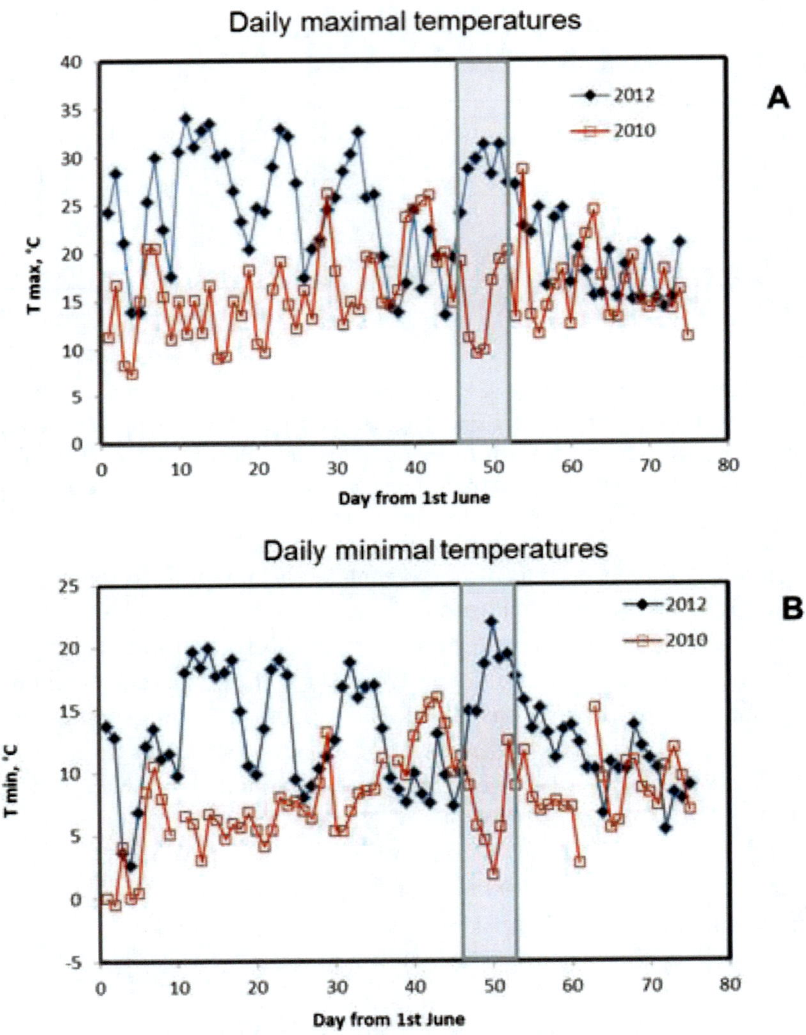

Figure 8. Plot of daily minimal and daily maximal temperatures at the closest to study site meteo station (Tarko-Sale) during « normal » summer 2010 and anomalously hot » summer 2012. The sampling period is highlighted in grey.Data source: http://www.mundomanz.com/.

Figure 9. Ground photos of shrank lakes and ponds during hot summer 2012 in the discontinuous permafrot zone of western Siberia.

Due to this extreme drought, the water levels in the large lakes decreased by c.a. 0.3-0.5 m. The small thermokarst lakes and ponds shrank significantly due to evaporation (Figure 9). During this drought, surface areas of most bodies of water (< 1000 m², ≤ 0.5 m depth) decreased by a factor between 2 and 5, and many small soil depressions and ponds completely disappeared, leaving dry bottoms covered by moss or organic-rich sediments (Figure 9 A, B). The anticyclone conditions persisted during the entire month of July in 2012, and the average surface water temperature (23.7±2.4° C, n = 40) remained approximately 10°C higher than during summer 2010 (14.1±2.2° C, Shirokova et al., 2013).

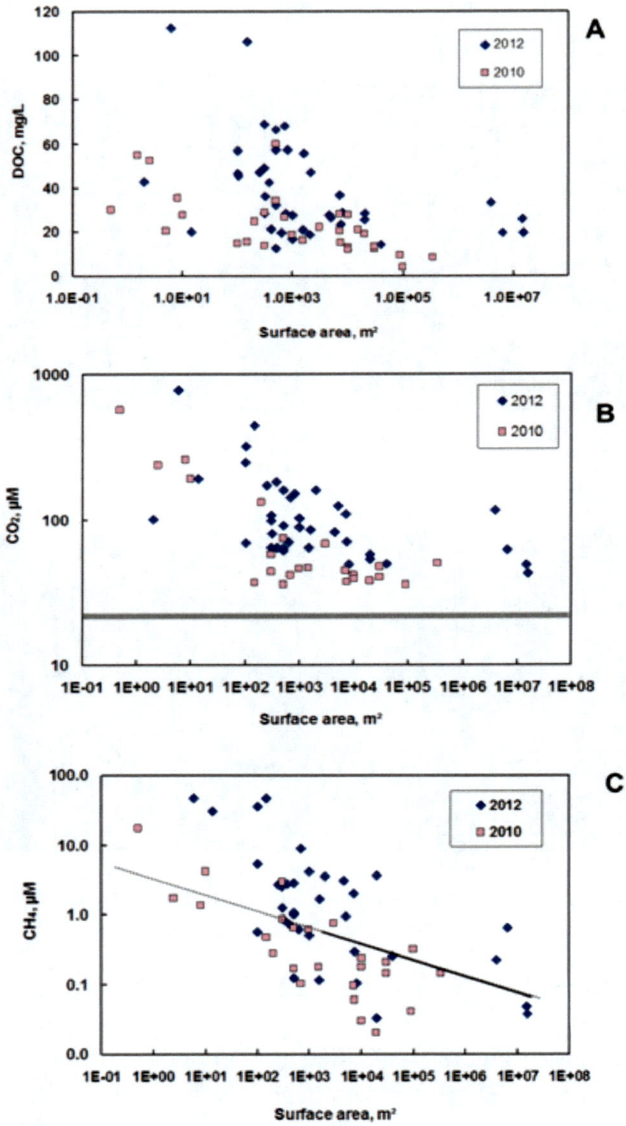

Figure 10. Plots of the concentrations of DOC (A), CO2 (B), and CH4(C) as a function of the lake body surface area in discontinuous permafrost zone sampled during normal summer 2010 (pink squares) and anomalously hot summer 2012. The grey line in Figure 10B represents equilibrium with the atmosphere and solid and dashed lines in Figure 10 C represent the data of Bastviken et al. (2004) and their extrapolation to the region of small surface areas.

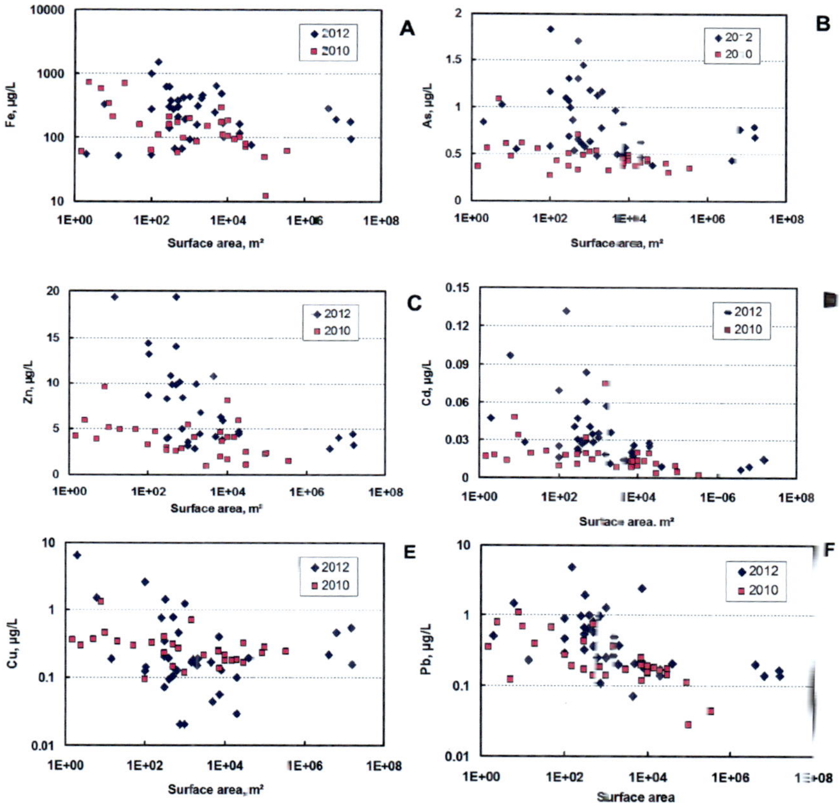

Figure 11. Plot of the concentration of Fe(A), As(B), Zn(C), Cd(D), Cu(E) and Cd(F) in thermokarst lakes of discontinuous / sporadic permafrost zone of western Siberia as a function of the water body surface area sampled during normal summer 2010 (pink squares) and anomalously hot summer 2012.

A plot describing the CO_2 and DOC relative to the water surface area (Figures 10 A & B, respectively) revealed that these forms of dissolved carbon increased their concentration in the water bodies with the surface area smaller than < 10^4 m² and remained quasi-constant in larger ponds and lakes with surface areas ranging from 10^4 to 10^7 m² . The increase in the CO_2 concentration spanned almost an order of magnitude in the smallest (< 10^2-10^3 m²) bodies of water compared to the larger ponds and lakes. The CO_2 concentration in the water samples during the summer of 2012 was 1.44 times higher than that in the normal summer recorded in 2010; moreover, the methane concentration in the summer of 2012 was 4.9 times higher than in

2010. For the sequence of thaw ponds and lakes that ranged from the local permafrost subsidences and shallow depressions to large, kilometer-sized lakes, we observed a systematic decrease in the total dissolved (< 0.45 μm) OC concentration relative to the water's surface area (Figure 10 B). The general trend observed for samples taken during the high temperature period was similar to the trend reported during the normal period. However, the mathematical and geometric averages of the DOC concentration across 40 ponds and thaw lakes during July 2012 were higher than during 2010 by 1.7 and 1.5 times, respectively.

Similar to carbon, most of the dissolved metals exhibited increased concentrations during summer 2012 over the entire range of thermokarst water bodies in the discontinuous permafrost region, as illustrated for Fe, As, Zn and Cd in Figure 11 (A-D). At the same time, Cu and Pb concentrations were not significantly affected by water heating (Figure 11 E-F).

Discussion and Implications for Permafrost Thawing and Water Warming

Intensification of Small Water Body Formation and Shortening the Life Cycle of Thermokarst Lakes

The first major consequence of the thawing permafrost in the central and northern areas of the Western Siberian lowlands is the acceleration of the thermokarst lakes' life cycle. The natural sequence of aquatic system development ("frozen mound – depression – small lakes – large lake – drained lake – khasyrey – palsa mound") will be affected such that there will be many small water bodies, but the mature, km-size lakes will be progressively drained, forming khasyreys bottoms (Kirpotin et al., 2007, 2008a, b; Dneprovskaya et al., 2009; Polischuk et al., this book). Consequently, the number of small permafrost subsidiences and thaw ponds will increase; these bodies are critical for mediating the GHG exchange with the atmosphere, as well as storing DOC and related dissolved metals storage in the water column.

Although the average increases in the thermokarst lake surface coverage has been 4-6 % during the last 36 years in the studied area (Kirpotin et al., 2008a, b), the decrease in the DOC concentration during lake maturation (a factor of 2 to 5) will certainly outweigh the surface area increase. Therefore, the total stock of DOC and dissolved metals in thermokarst lakes will progressively decrease. This decrease will be accompanied by a relative increase in the aquagenic OM (water column-derived biopolymers and polysaccharides) versus the pedogenic fulvic acids, most likely following the same trend as was observed for the temperate lakes (from April to July, Wilkinson et al., 1997).

The typical warming scenario in the Yukon River and other Artic river basins suggests that the export flux of terrestrial OC into the ocean will increase because the old OC stored in the Arctic peatlands and permafrost will be transported as particulate material (Guo and MacDonald, 2006) with low bioavailability (Benner et al., 2004). Small rivers may deliver their particulate OM load to newly formed thermokarst lakes, where it will be buried in the sediments instead of entering the hydrological network. In contrast to the rivers, the thermokarst lakes' proportion of the particulate phase is likely to be small; most of the OC present in the colloidal state is vulnerable to significant bacterial and photodegradation. These degradation processes will likely intensify under warmer conditions. Therefore, the organic carbon present in the thermokarst lakes may be much more reactive and susceptible toward exchange with the atmosphere versus the OC in rivers and soils, increasing the positive feedback for permafrost systems' response to the climate change.

Overall, the thermokarst lakes in western Siberia are an important site for degrading the released terrestrial permafrost OC. Further studies regarding the lakes' biogeochemistry in the region of discontinuous and permafrost-free environments in western Siberia are needed to predict the net response of the boreal zone toward the on-going environmental changes.

Substituting "Space for Time" Using Latitude Gradient of Thermokarst Lakes

A straightforward method for predicting the thermokarst lake water changes under climate warming scenario is "substituting space for time," as previously used in western Siberia for the rivers (Frey et al. 2007; Frey and McClelland, 2009). This model assumes that the contemporary difference on the local and global scale within the climate gradient created by latitude or by slope orientation can serve as a proxy for the future changes to a given system. The permafrost degradation that leads to a shift from continuous to discontinuous and sporadic distribution of permafrost will be the most important factor that controls the dissolved metal, DOC and GHG regime in the thermokarst lakes' water column.

Analyzing the measured elemental concentrations relative to the latitude revealed several patterns, as illustrated in Figure 7 A-H. These patterns might reflect the superimposition of various processes, such as the intensity of element leaching from peat, feeding by the groundwater, mineralization in the water column and atmospheric deposition of marine aerosols. Therefore, leaching the DOC and TE from the frozen peat should be efficient in the south, within the sporadic/discontinuous permafrost zone, due to the elevated water temperatures and thicker unfrozen (active) layer of soil. As such, the dissolved components from the peat should decrease in concentration in the northward direction. This is especially true for Pb (Figure 7 G) known to be strongly concentrated in peat deposits (Shotyk et al., 2000). Concurrently, the DOC and related organo-mineral (Fe, Al) colloids are respired by aerobic heterotrophic bacterioplankton in western Siberia thermokarst water bodies (Shirokova et al., 2013). The activity of bacterioplankton is likely to increase with the water temperature increase. The superimposition of the two main processes leading to the DOC and TE release and uptake in the water column may cause the complex, convex-like pattern of concentration versus latitude, as displayed in Figure 7 for Fe and Al. A similar convex pattern is observed for Mn (not shown); its maximum concentration occurs at 66°N. For this element, the photo-oxidation processes increase in the northward direction and can compete with the heterotrophic respiration that increases in the opposite direction. Both processes can transform the dissolved (< 0.22 µm) fraction of an element into a particulate phase that accumulates in the sediment. The last factor that may affect the metal concentration is the proximity of the Norilsk Ni-Cu-Cr-Pd smelter; this facility is the largest metal sulfide processing factory in the Arctic. The increased Ni, Cr and Cu concentrations in the lake water on the northernmost site of the Gyda peninsula (Figure 7 F, H) may reflect the atmospheric deposition of metal-rich aerosols that originate from the Norilsk smelters located 350 km SE from the site.

Natural Climate Perturbation As an Approach to Model the Future Climate Change

Local weather perturbations persisting for more than month are sufficient to significantly affect the processes occurring in the thermokarst lakes; these extreme weather effects offer a unique opportunity to directly test the effects of warmer water on the lakes' chemical composition and the CO_2 exchange with the atmosphere. Climate warming at high latitudes triggers enhanced

GHG emissions. Therefore, to predict future scenarios involving a warmer and drier climate (Porcal et al., 2009), the lake water levels may fall and expose previously stored organic sediments to greater aeration, releasing more CO_2 into the atmosphere. The relative increase in the aqueous GHG concentrations during the anomalously hot summer of 2012 revealed the northern permafrost–controlled aquatic systems' vulnerability. The concentrations of the major and trace inorganic elements increased 2 to 5 times during the summer of 2012 relative to the normal summer of 2010. These results for inorganic elements exceed the conservative scenario linked merely to water evaporation.

There are 4 major processes that are likely to control the degree and direction of the aquatic ecosystems' response to prolonged warming in the permafrost region's water column in the. Evaporative concentration can increase the DOC and dissolved metal concentrations 2-fold, considering the shallowness of the studied thaw ponds and lakes. For dissolved CO_2 and CH_4, however, their concentrations are determined by chemical and microbiological processes that occur in the sediments and in the water column. Because the studied water bodies receive no input from the surface streams that drain the adjacent soil and groundwater, accelerated soil respiration or plant growth caused by global warming should not directly affect the thermokarst lakes' chemical composition. In contrast, for CO_2, the heterotrophic respiration of DOM in the water column and the subsequent CO_2 production in the sediments should both respond positively to the rising water temperatures. The impact on methane concentration will be even stronger due to the decrease in the oxygenated waters' depth, the enhanced O_2 uptake by heterotrophic respiration in the water column and the intensified methanogenesis in the sediments. For DOC, the major cations and trace metals involved in the increased C_{org} (in the form of DOM) leaching rate from the peat outcroppings at the lake borders may increase of the concentration 3-fold for 10°C temperature increase due to the typical activation energy of the solid phase's dissolution in water. Consequently, in the Khanymey lakes studied in this work, heterotrophic respiration of dissolved organic carbon could not overcome enhanced peat leaching and physical evaporation despite almost a 10°C increase in the water temperature during more than 2 months of drought; leaching and evaporation are more important factors that generate more DOC in ecosystems with higher CO_2 and water levels during short-term climate warming scenarios. Therefore, various processes may occur during surface water warming that lead to the non-linear response observed in the thermokarst water bodies' chemical composition toward water evaporation. Based on our field observations and available analytical data, we list these processes in the decreasing order of their impact: leaching of soluble peat components at the lake border > heterotrophic mineralization of DOM by aerobic bacterioplankton in the water column > photodegradation of DOM in the surface layer of the water body. This order reflects the water temperature's effect on the increasing DOC, CO_2 and metal concentrations during water warming. The fate of trace metals in the thermokarst lakes after surface water warming is an important and understudied issue. From a recent speciation and size fractionation study in these lakes, the metal bioavailability may rise in response to the temperature increases caused by the decrease in the proportion of colloidal material (1 kDa – 0.22 μm) and the increase of the fully dissolved low molecular weight (< 1-2 nm) fraction in the water column (Pokrovsky et al., 2013). Currently, it remains unclear whether the trace elements limit this region given that P and N typically limit the primary production in freshwater systems. The thermokarst lake systems are different from many other continental water settings. First, the systems examined in this study cannot be considered similar to other freshwater systems: these are dystrophic, acidic waters, with high DOC and very low total dissolved solid concentrations. The km-sized permanent lakes have a 10 to 15 μS cm^{-1}specific conductance; this value is an order of magnitude lower than that in other known pristine freshwater systems. Under these conditions, N, P and trace metals, such as Fe, Mn, Zn, Co, may become limiting materials. Second, there is virtually no phytoplankton or primary productivity in studied water bodies, including the large thermokarst lakes (Shirokova et al., 2009, 2013). The limitation of heterotrophic bacterioplankton by trace metals is currently unknown. Finally, similar to the oligotrophic coastal waters, we cannot exclude the possibility that metal micronutrients, such as Fe, may become limiting at certain conditions in the thermokarst lakes. Regardless of the trace metal limitation mechanisms for aquatic biota in the thermokarst water bodies, the increase in the low molecular weight fraction of the dissolved trace metal may increase the delivery of this fraction to the rivers and eventually the Arctic Ocean, therefore modifying the response of the primary producers in the permafrost zone due to climate change.

CONCLUSION

Thermokarst water bodies form within the natural succession of frozen palsa bog thawing and permafrost subsidiences; these bodies are highly dynamic systems that reflect the ongoing climate and environmental changes in the chemical composition of their water columns. The initial permafrost thawing in the palsa frozen peat bog complex creates reactive sites for GHG exchange between the frozen organic matter and the atmosphere. Although the exact number and the degree of land surface coverage by these small (< 100 m^2) water bodies is currently unknown, their role in the overall CO_2 and CH_4 flux from the land to the atmosphere may be higher than that of the large thermokarst lakes or land bog complexes.

Examining the thaw lake chemical composition along the latitudinal profile of variable permafrost thickness and coverage demonstrated that the average DOC, Pb, Al concentration decreased, while the specific conductance increased with the Ca, Ni, Cu concentrations from the south to the north; these changes are most likely linked to the permafrost's variability and the effects caused by the sea or ore processing smelter's proximity.

Natural weather perturbation phenomena, such as prolonged water heating during the summer, allowed us to account for various physico-chemical and microbiological processes that control the organic carbon's leaching from the frozen peat, as well as its transformation in the water column and the release to the atmosphere. There is little doubt that as the climate change in the Western Siberia lowlands continues, the thermokarst thaw ponds will govern the GHG released into the atmosphere.

Acknowledgments

The financial support from a BIO-GEO-CLIM Mega-grant from the Ministry of Education and Science of the Russian Federation and Tomsk State University (No 14.B25.31.0001) and ANR "Arctic Metals" are acknowledged.

References

Bastviken, D., Cole, J., Pace, M., Tranvik, L., 2004. Methane emissions from lakes: Dependence on lake characteristics, two regional assessments, and a global estimate. Global Biogeochem Cycles, 18: GB4009, doi:10.1029/2004GB002238.

Benner, R., Benitez-Nelson, B., Kaiser, K., Amon, R.M.W., 2004. Export of young terrigenous dissolved organic carbon from rivers to the Arctic Ocean. *Geophys. Res. Lett.*, 31, L05305, doi: 10.1029/2003GL019251.

Christensen T.R., Johansson T., Akerman H.J., Mastepanov M., Malmer N., Friborg T., Crill P., Svensson, B.H., 2004. Thawing sub-arctic permafrost: Effects on vegetation and methane emissions. *Geophys. Res. Letters*, 31, L04501, doi: 10.1029/2003GL018680.

Dneprovskaya, V.P., Bryksina, N.A., Polischuk, Yu.M., 2009. Study of thermokarst changes in discontinuous zone of west-Siberian permafrost based on space images. *Study of the Earth from the Space (Issledovanie Zemli is Kosmosa)*, No 4, 88-96.

Frey, K. E., Siegel, D.I., Smith, L.C., 2007. Geochemistry of west Siberian streams and their potential response to permafrost degradation. *Water Resour. Res.*, 43, W03406, doi: 10.1029/2006WR004902.

Frey, K.E., McClelland, J.W., 2009. Impacts of permafrost degradation on arctic river biogeochemistry. *Hydrological Process.*, 23, 169–182.

Heikkinen, J.E.P., Virtanen, T., Huttunen, J.T., Elsakov, V., Martikainen, P.J., 2004. Carbon balance in east European tundra. *Global Biogeochem. Cycles*, 18, GB1023, doi: 10.1029/2003GB002054.

Hinkel, K.M., Eisner, W.R., Bockheim, J.G., Nelson, F.E., Peterson, K.M., Dai, X., 2003. Spatial extent, age, and carbon stoks in drained thaw lake basins on the Barrow Peninsula, Alaska. *Arctic, Antarctic, and Alpine Research*, 35, 291-300.

Jorgenson, M.T., Racine, C.H., Walters, J.C., Osterkamp, T.E., 2001. Permafrost degradation and ecological changes associated with a warming in central Alaska. *Clim. Change* 48, 551-579.

Kirpotin, S.N., Naumov, A.V., Vorobiov, S.N., Mironycheva-Tokareva, N.P., Kosych, N.P., Lapshina, E.D., Marquand, J., Kulizhski, S.P., Bleuten, W., 2007. Western-Siberian Peatlands: Indicators of Climate Change and Their Role in Global Carbon Balance. Chapter 33 in Climate Change and Terrestrial Carbon Sequestration in Central Asia / edited by R. Lal, M. Suleimenov, B.A. Stewart, D.O. Hansen, and P. Doraiswamy, Taylor and Francis, Amsterdam, Holland, pp. 453-472.

Kirpotin, S.N., Polishchuk, Yu.M., Bryksina, N.A., 2008a. Dynamics of thermokarst lakes areas in continuous and discontinuous cryolithozones of Western Siberia under global warming. *Vestnik of Tomsk State University*, 311, 185-189 (in Russian).

Kirpotin, S., Polishchuk, Yu., Zakharova, E., Shirokova, L., Pokrovsky, O., Kolmakova, M., Dupre, B., 2008b. One of possible mechanisms of thermokarst lakes drainage in West-Siberian North. *Internat. J. Environ. Studies*, 65, 631-635.

Kirpotin, S.N., Berezin, A., Bazanov, V., Polishchuk, Y., Vorobiev, S., Mozgolin, B., Aakerman, E., Mironycheva-Tokareva, N., Volkova, I., Dupré, B., Porkovsky, O.S., Koiraev, A., Zakharova, E., Shirokova, L., Viers, J., Kolmakova, M., 2009. West Siberian wetlands as indicator and regulator of climate change on the global scale. *Internat. J. Environ. Studies*, 66(4), 409-421.

Kling, G.W., Kipphut, G.W., Miller, M.C., 1992. The flux of CO_2 and CH_4 from lakes and rivers in arctic Alaska. *Hydrobiologia*, 240, 23-36.

Kortelainen, P., Rantakari, M., Huttunen, J.T., Mattsson, T., Alm, J., Juutinen, S., Larmola, T., Silvola, J., Martikainen, P.J., 2006. Sediment respiration and lake trophic state are important predictors of large CO_2 evasion from small boreal lakes. *Glob. Change Biol.*, 12, 1554–1557.

Laurion, I., Vincent W.F., MacIntyre, S., Retamal, L., Dupont, C., Francus, P., Pienitz, R., 2010. Variability in greenhouse gas emissions from permafrost thaw ponds. *Limnol. Oceanogr.*, 55, 115-133.

Payette, S., Delwaide, A., Caccianiga, M., Beauchemin, M., 2004. Accelerated thawing of subarctic permafrost over the last 50 years. *Geophys. Res. Letters*, 31, L18208, doi: 10.1029/2004GL020358.

Pokrovsky, O.S., Schott, J., 2002. Iron colloids/organic matter associated transport of major and trace elements in small boreal rivers and their estuaries (NW Russia). *Chem. Geol.*, 190, 141-179

Pokrovsky, O. S., Schott, J., Dupré, B., 2006. Trace element fractionation and transport in boreal rivers and soil porewaters of permafrost-dominated basic terrain in Central Siberia. *Geochim. Cosmochim. Acta*, 70, 3239-3260.

Pokrovsky, O.S., Viers, J., Shirokova, L.S., Shevchenko, V.P., Filipov, A.S., Dupré, B., 2010. Dissolved, suspended, and colloidal fluxes of organic carbon, major and trace elements in Severnaya Dvina River and its tributary. *Chem. Geol.*, 273, 136–149.

Pokrovsky, O.S., Shirokova, L.S., Kirpotin, S.N., Audry, S., Viers, J., Dupré, B., 2011. Effect of permafrost thawing on the organic carbon and metal speciation in thermokarst lakes of western Siberia. *Biogeosciences*, 8, 565–583.

Pokrovsky, O.S., Shirokova, L.S., Kirpotin, S.N., Kulizhsky, S.P., Vorobiev, S.N., 2013. Effects of anomalous high temperatures on carbon dioxide, methane, dissolved organic carbon and trace element concentrations in thaw lakes in Western Siberia in 2012. *Biogeosciences Discussion*, 10, 7257-7297, doi: 10.5194/bgd-10-7257-2013.

Porcal, P., Koprivnjak, J.-F., Molot, L.A., Dillon, P.J., 2009. Humic substaces – part 7: the biogeochemistry of dissolved organic carbon and its interactions with climate change. *Environ. Sci. Pollut. Res.*, 16, 714–726.

Ramlal, P.S., Hesslein, R.H., Hecky, R.E., Fee, E.J., Rudd, J.W.M., Guildford, S.J., 1994. The organic carbon budget of a shallow Arctic tundra lake on the Tuktoyaktuk Peninsula, N.W.T., Canada. *Biogeochemistry*, 24, 145-172.

Riordan, B., Verbyla, D., McGuire, A.D., 2006. Shrinking ponds in subarctic Alaska based on 1950-2002 remotely sensed images. *J. Geophys. Res.*, 11, G04002, doi: 10.1029/2005JG000150.

Ripo, M.E., Huttunen, J.T., Naumov, A.V., Chichulin, A.V., Lapshina, E.D., Bleuten, W., Martikainen, P.J., 2007. Release of CO_2 and CH_4 from small wetlands lakes in western Siberia. *Tellus*, 59B, 788–796.

Schuur, E.A.G., Bockhein, J., Canadell, J.P., Euskirchen, E., Field, C.B., Goryachkin, S.V., Hagemann, S., Kuhry, P., Lafleur, P.M., Lee, H., Mazhitova, G., Nelson, F.E., Rinke, A.,

Romanovsky, V.E., Shiklomanov, N., Tarnocai, C., Venesy, S., Vogel, J.G., Zimov, S.A., 2008. Vulnerability of permafrost carbon to climate change: Implications for the global carbon cycle. *BioScience*, 58, 701-714.

Shirokova, L.S., Pokrovsky, O.S., Kirpotin, S.N., Dupré, B., 2009. Heterotrophic bacterio-plankton in thawed lakes of northern part of Western Siberia controls the CO_2 flux to the atmosphere. *Internat. J. Environ. Stud.*, 66, N 4, 433-445.

Shirokova, L.S., Pokrovsky, O.S., Viers, J., Klimov, S.I., Moreva, O.Yu., Zabelina, S.A., Vorobieva, T.Ya., Dupré, B., 2010. Diurnal variations of trace elements and heterotrophic bacterioplankton concentration in a small boreal lake of the White Sea basin. *Ann. Limnol. - Int. J. Lim.*, 46, 67–75.

Shirokova L.S., Pokrovsky O.S., Kirpotin S.N., Desmukh C., Pokrovsky B.G., Audry S., Viers J., 2013. Biogeochemistry of organic carbon, CO_2, CH_4, and trace elements in thermokarst water bodies in discontinuous permafrost zones of Western Siberia. *Biogeochemistry*, 113, 573-593.

Shotyk, W., Blaser, P., Grünig, A., and Cheburkin, A. K., 2000. A new approach for quantifying cumulative, anthropogenic, atmospheric lead deposition using peat cores from bogs: Pb in eight Swiss peat bog profiles. Sci. Total Environ., 249, 281–295.

Turetsky M.R., Wieder R.K., Vitt D.H., 2002. Boreal peatland C fluxes under varying permafrost regimes. *Soil Biol. Biochem.*, 34, 907-912.

Vasyukova, E.V., Pokrovsky, O.S., Viers, J., Oliva, P., Dupré, B., Martin, F., Candaudaup, F., 2010. Trace elements in organic- and iron-rich surficial fluids of the Boreal zone: Assessing colloidal forms via dialysis and ultrafiltration. *Geochim. Cosmochim. Acta*, 74, 449-468.

Wickland K.P., Striegl R.G., Neff J.C., Sachs Th., 2006. Effects of permafrost melting on CO_2 and CH_4 exchange of a poorly drained black spruce lowland. *J. Geophys. Res.* 111, G02011, doi: 10.1029/2005JG000099.

Wilkinson, K.J., Joz-Roland, A., Buffle, J., 1997. Different roles of pedogenic fulvic acids and aquagenic biopolymers on colloid aggregation and stability in freshwaters. *Limnol. Oceanogr.,* 42, 1714-1724.

Zimov, S.A., Voropaev, Y.V., Semiletov, I.P., Davidov, S.P., Prosiannikov, S.F., Chapin III, F.S., Chapin, M.C., Trumbore, S., Tyler, S., 1997. North Siberian lakes: a methane source fueled by Pleistocene Carbon. *Science*, 277, 800-802.

Zakharova, E.A., Kouraev, A.V., Kolmakova, M.V., Mognard, N.M., Zemtsov, V.A., Kirpotin, S.N., 2009. The modern hydrological regime of the northern part of Western Siberia from in situ and satellite observations. *Internat. J. Environ. Stud.*, 66, 447-463.

In: Permafrost
Editor: Oleg S. Pokrovsky
ISBN: 978-1-62948-830-1

Chapter 8

PENICILLIUM FUNGI FROM PERMAFROST: BIOSYNTHESIS OF SECONDARY METABOLITES, PECULIARITIES OF GROWTH AND DEVELOPMENT

A. G. Kozlovsky, V. P. Zhelifonova and T. V. Antipova
G.K. Skryabin Institute of Biochemistry and Physiology
of Microorganisms, Russian Academy of Sciences, Russia

ABSTRACT

Secondary metabolite biosynthesis was studied in the strains of *Penicillium* fungi isolated from permafrost. For strains isolated from permafrost, a low speed of growth is characteristic. Secondary metabolites were found to be synthesised by 59% of the permafrost strains studied; 25 known secondary metabolites were identified and the structures of two new compounds, quinocitrinines, were elucidated. The cultures are producers of various structural groups of secondary metabolites. Clavine alkaloids were represented by α-cyclopiazonic acid, fumigaclavines A and B, festuclavine, rugulovasines A and B, agroclavine-1, and epoxyagroclavine-1. Diketopiperazine alkaloids of the roquefortine group were represented by tryptophanyl dehydrohistidyl diketopiperazine, 3,12-dihydroroquefortine, roquefortine, glandicolines A and B, and meleagrine. Benzodiazepine alkaloids cyclopeptine and cyclopenine were also identified. Quinoline alkaloids were represented by viridicatine and quinocitrinines; and quinozaline alkaloids, by fumiquinozalines F and G. Substances of polyketide nature were identified: ochratoxins A and B, mycophenolic acid, griseofulvin, and PC-2. No principal differences exist in the profiles of secondary metabolites for strains isolated from permafrost and from a usual habitat.

INTRODUCTION

The microbial ecology and biodiversity of permafrost have been extensively reviewed (Wagner, 2008). Viable microfungi and bacteria were reported for Arctic and Antarctic permafrost sediments of different ages using cultural and molecular techniques (Gilichinsky et al., 2007; Kochkina et al., 2001; Ozerskaya et al., 2009; Zucconi et al., 2012). Recent data suggest considering the permafrost environment as an active microbial ecosystem rather than a frozen habitat with microbial survivors. The stability of the environmental parameters below the superficial layer makes this environment less extreme than one supposes and it may be ideal for the long-term preservation of microorganisms and biomolecules (Willerslev et al., 2004).

Penicillium fungi are known to be promising subjects for the search of novel biologically active compounds (Dreyfuss and Chapela, 1994). They are producers of various types of compounds such as ergot alkaloids, diketopiperazines, quinolines, quinazolines, benzodiazepines, and polyketides.

Species of the genus *Penicillium* are considered difficult to identify using conventional microbiological approaches. The general identification of *Penicillium* fungi by micro- and macromorphological features often gives no definite results. The reliability of *Penicillium* fungus classifications to certain species seems possible due to the species-specific production of various biologically active compounds (Samson and Frisvad, 2004). Penicilles isolated from long-frosted deposits can be expected with high probability to synthesise secondary metabolites, potential biologically active compounds, which can help the producer to survive and adapt to given conditions.

MORPHOLOGICAL, PHYSIOLOGICAL, AND BIOCHEMICAL PECULIARITIES OF *PENICILLIUM* ISOLATED FROM PERMAFROST

Various probes of permafrost were found to contain predominantly mycelial fungi, *Penicillium* spp., including small-sized single-celled conidia

(Ozerskaya et al., 2009). Low temperature was shown to stimulate protective mechanisms peculiar to fungal cells. These mechanisms include increased concentrations of trehalose, polyols, and polyunsaturated fatty acids inside the cells, as well as the synthesis of enzymes active at low temperatures (Robinson, 2001).

These mechanisms are not only conducive to the survival of fungi under conditions of low temperature, low water activity, and hypoxia, but also contribute to the development of some strains in certain species under these extreme conditions. Therefore, *Penicillium* fungi isolated from long-frosted deposits can be expected to synthesise unknown secondary metabolites, potential biologically active compounds, which can help the producer to survive and adapt to these conditions.

Penicillium fungi isolated from permafrost are characterised by high species diversity and a high frequency of occurrence. Various biotopes of permafrost have been found to contain penicilles that are also widespread in various modern habitats: *P. aurantiogriseum*, *P. brevicompactum*, *P. chrysogenum*, *P. crustosum*, *P. granulatum*, *P. griseofulvum*, *P. puberulum*, *P. verrucosum*, *P. viridicatum*, *P. citrinum*, *P. melinii*, *P. miczynskii*, *P. simplicissimum*, *P. decumbens*, *P. glabrum*, *P. restrictum*, *P. miniolutеum*, *P. purpurogenum*, *P. rugulosum*, and *P. variabile* (Ozerskaya et al., 2009). These species are highly adaptable to the environment and are capable of colonising diverse organic matter; many species are halotolerant and psychrophilic (Pitt, 1979; Samson and Frisvad, 2004).

PENICILLIUM SECONDARY METABOLITES FROM PERMAFROST

The biosynthesis of secondary metabolites was studied in 83 *Penicillium* strains (Antipova et al., 2008; Antipova et al., 2011; Zhelifonova et al., 2006; Zhelifonova et al., 2009; Kozlovsky et al., 2003a; Kozlovsky et al., 2003b; Kozlovsky et al., 2003c; Kozlovsky et al., 2009, Kozlovsky et al., 2012) obtained from the All-Russian Collection of Microorganisms (VKM). The strains were isolated from soils, water, cryopegs, and volcanic ashes of various geologic ages sampled in various permafrost regions: Antarctica, North America (Canada), North-East Asia (Kolyma), and Kamchatka (Ozerskaya et al., 2004; Ozerskaya et al., 2009; Kochkina et al., 2001; Gilichinsky et al., 2005; Gilichinsky et al., 2007).

Table 1. Secondary metabolite profiles and taxonomic positions of the strains isolated from permafrost

Species	Strain number VKM FW	Identified secondary metabolites	Age of sample, thousand years	Isolation site
Subgenus *Penicillium*				
P. brevi-compactum	791	Mycophenolic acid	Holocene	Antarctica, soil
	725		20–30	Arctic, Kolyma, soil
P. chysogenum	2877	Roquefortine	30	Antarctica, soil
	2873	Roquefortine, meleagrine	15–40	Arctic, Kolyma, soil
	2835	Roquefortine, 3,12-dihydroroquefortine	15–40	Arctic, Kolyma, soil
	2739	Roquefortine, 3,12-dihydroroquefortine, tryptophanyl dehydrohistidyl diketopiperazine	15–40	Arctic, Kolyma, soil
	2863	Roquefortine, 3,12-dihydroroquefortine, tryptophanyl dehydrohistidyl diketopiperazine, glandicolines A and B	–*	Kamchatka, volcanic ashes
	653	Not identified	170	Arctic, Canada, soil
	659		170	Arctic, Canada, soil
	679		100	Arctic, Kolyma, soil
	684		100	Arctic, Kolyma, soil
	694		100	Arctic, Kolyma, soil
	721		20–30	Arctic, Kolyma, soil
	778		5–10	Arctic, Kolyma, soil
	799		1800–3000	Arctic, Kolyma, soil
	812		1800–3000	Arctic, Kolyma, soil
P. commune	2885	α-Cyclopiazonic acid (CPA)	150	Antarctica, soil
	2881		150	Antarctica, soil
	2851		1800–3000	Arctic, Kolyma, soil
	2830		1800–3000	Arctic, Kolyma, soil
	2829		15–40	Arctic, Kolyma, soil
	1447		25–40	Arctic, Kolyma, soil
	2876		15–40	Arctic, Kolyma, soil
	2753		100	Arctic, Kolyma, cryopeg
	2666		–	Kamchatka, volcanic ashes
	2853		–	Kamchatka, volcanic ashes
P. griseofulvum	2251	Griseofulvin	600-1800	Arctic, Kolyma, soil
P. melano-conidium	738	Roquefortine, 3,12-dihydroroquefortine	30	Antarctica, soil
	741		30	Antarctica, soil
	766		3000	Arctic, Kolyma, soil

Species	Strain number VKM FW	Identified secondary metabolites	Age of sample, thousand years	Isolation site
P. palitans	2852	CPA, fumigaclavines A and B, festuclavine	–	Kamchatka, volcanic ashes
	2854		–	Kamchatka, volcanic ashes
	794	Fumigaclavines A and B, festuclavine	Holocene	Antarctica, soil
	657		170	Arctic, Canada, soil
	667		170	Arctic, Canada, soil
	690		100	Arctic, Canada, soil
	704		100	Arctic, Canada, soil
P. solitum	2921	Cyclopenine, cyclopeptine	Modern	Antarctica, water
	2928	Cyclopenine, cyclopeptine, viridicatine	ditto	Antarctica, water
	2600	Cyclopenine, cyclopeptine	3000	Arctic, Kolyma, soil
	2604	Cyclopenine, cyclopeptine, viridicatine	ditto	Arctic, Kolyma, soil
	2615	Cyclopenine, cyclopeptine, cyclopenol	15	Arctic, Kolyma soil
	2665	Cyclopenine, cyclopeptine	–	Kamchatka, volcanic ashes
P. thymicola	2648	Fumiquinazolines F and G, PC-2	Modern	Arctic, Kolyma, soil
	869		100–120	Arctic, Kolyma, cryopeg
P. verrucosum	875	Ochratoxins A and B	100–120	Arctic, Kolyma, water-suspended matter
	877		ditto	Arctic, Kolyma, water-suspended matter
	878		ditto	Arctic, Kolyma, water-suspended matter
	907		ditto	Arctic, Kolyma, cryopeg
	908		ditto	Arctic, Kolyma, cryopeg
P. viridicatum	809	Not identified	1800–3000	Arctic, Kolyma, soil
	2859		150	Antarctica, soil
Subgenus *Furcatum*				
P. citrinum	800	Agroclavine-1, epoxyagroclavine-1, quinocitrinines A and B	1800–3000	Arctic, Kolyma, soil
	2621		Modern	Arctic, Kolyma, soil
	2629		Modern	Arctic, Kolyma, soil
	2631		Modern	Arctic, Kolyma, soil
	2882		150	Antarctica, soil

Table 1. (Continued)

Species	Strain number VKM FW	Identified secondary metabolites	Age of sample, thousand years	Isolation site
P. simplicissimum	2435	Not identified	15–40	Arctic, Kolyma, soil
	2461		5–10	Arctic, Kolyma, soil
	2735		5–10	Arctic, Kolyma, soil
P. waksmanii	2875	Agroclavine-1, epoxyagroclavine-1, quinocitrinines A and B	100–200	Arctic, Kolyma, soil
	2866	Not identified	1800–3000	Arctic, Kolyma, soil
	2936		–	Antarctica, water
Subgenus *Biverticillium*				
P. minioluteum	661	Not identified	170	Arctic, Canada, soil
	773		5–10	Arctic, Kolyma, soil
	782		5–10	Arctic, Kolyma, soil
P. purpurogenum	772	Not identified	5–10	Arctic, Kolyma, soil
P. rugulosum	665	Not identified	170	Arctic, Canada, soil
	717		20–30	Arctic, Kolyma, soil
	733		30	Antarctica, soil
	769		3000	Arctic, Kolyma, soil
P. variabile	655	Rugulovasines A and B	170	Arctic, Canada, soil
	806		1800–3000	Arctic, Kolyma, soil
	811		ditto	Arctic, Kolyma, soil
	816		ditto	Arctic, Kolyma, soil
	818		ditto	Arctic, Kolyma, soil
	899	Not identified	100–120	Arctic, Kolyma, water-suspended matter
	1448		25–40	Arctic, Kolyma, soil
	1449		1800–3000	Arctic, Kolyma, soil
	1451		5–10	Arctic, Kolyma, soil
	1452		15–30	Arctic, Kolyma, soil
	2884		150	Antarctica, soil
	2886		150	Antarctica, soil
Subgenus *Aspergilloides*				
P. decumbens	689	Not identified	100	Arctic, Kolyma, soil
	692		100	Arctic, Kolyma, soil
	693		100	Arctic, Kolyma, soil
P. glabrum	652	Not identified	170	Arctic, Canada, soil
	678		100	Arctic, Kolyma, soil

* indicates no date.

Secondary metabolites were found to be synthesised by 59% of the permafrost strains studied; 25 known secondary metabolites were identified and the structures of two new compounds, quinocitrinines, were elucidated (Table 1). The cultures are producers of various structural groups of secondary metabolites. Clavine alkaloids were represented by α-cyclopiazonic acid (CPA), fumigaclavines A and B, festuclavine, rugulovasines A and B, agroclavine-1, and epoxyagroclavine-1 (Figure 1). Diketopiperazine alkaloids of the roquefortine group were represented by tryptophanyl dehydrohistidyl diketopiperazine, 3,12-dihydroroquefortine, roquefortine, glandicolines A and B, and meleagrine (Figure 2).

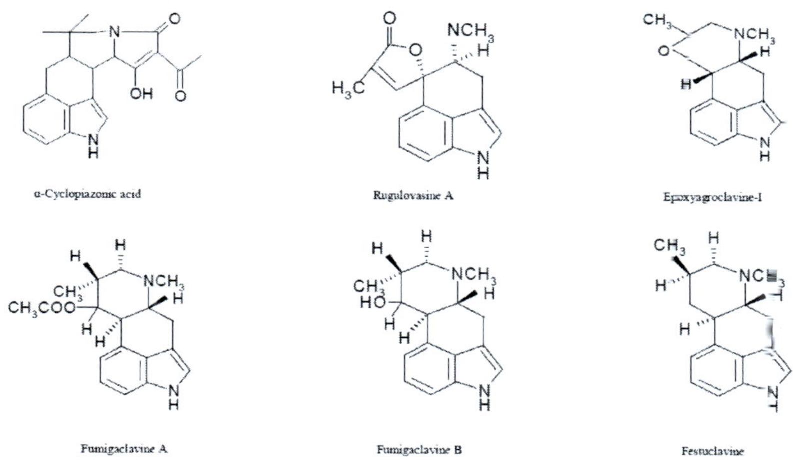

Figure 1. Structures of clavine ergot alkaloids.

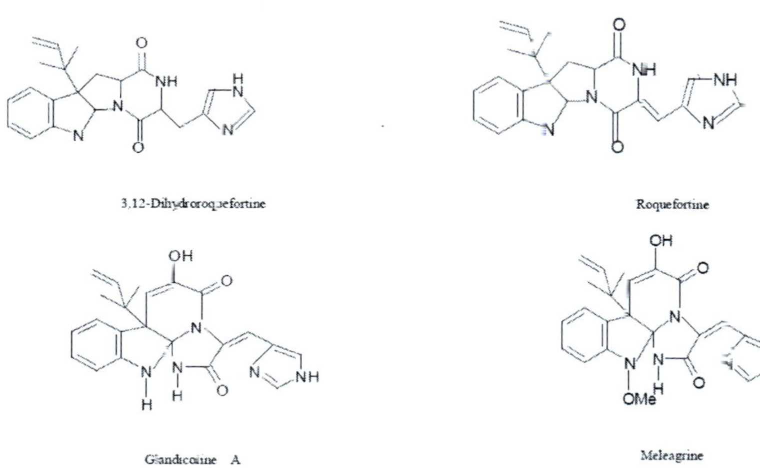

Figure 2. Structures of diketopiperazine alkaloids.

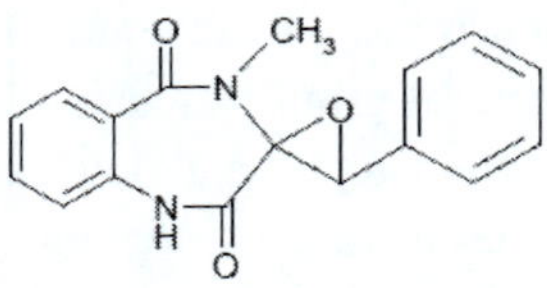

Cyclopenine

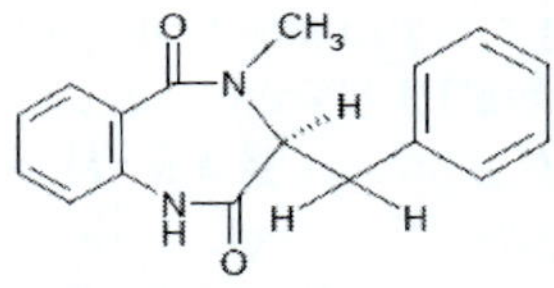

Cyclopeptine

Figure 3. Structures of benzodiazepine alkaloids.

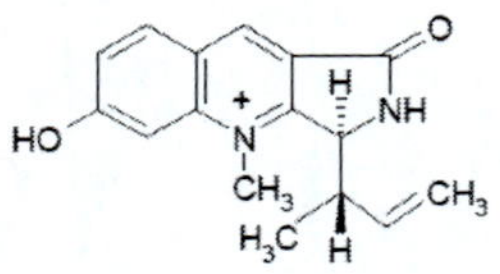

Quinocitrinine A

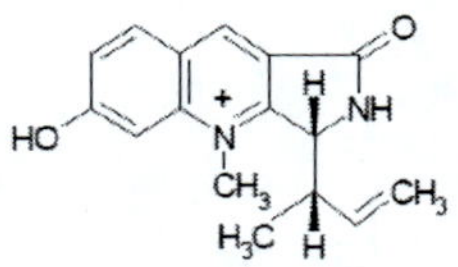

Quinocitrinine B

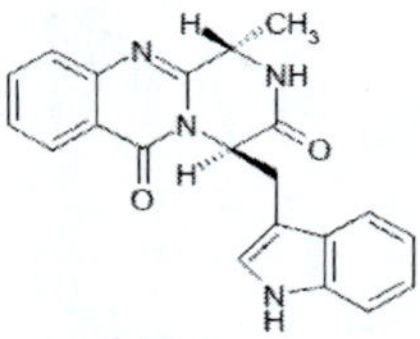

Fumiquinazoline F

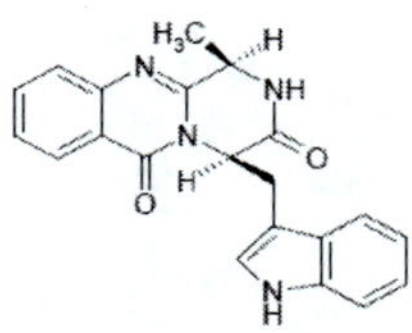

Fumiquinazoline G

Figure 4. Structures of quinoline and quinazoline alkaloids.

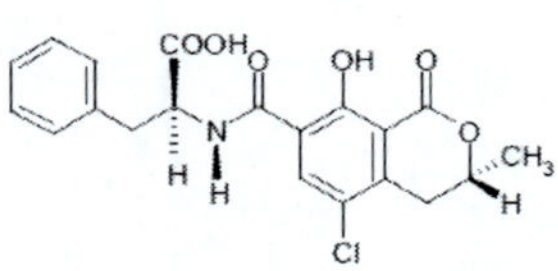

Ochratoxin A

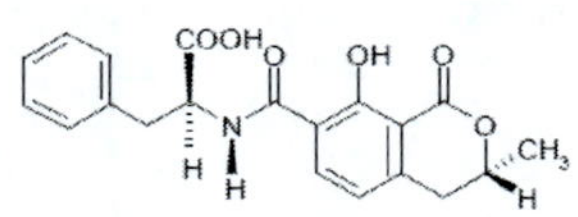

Ochratoxin B

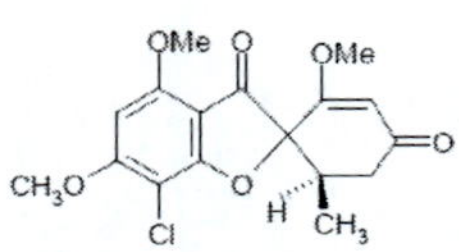

Griseofulvin

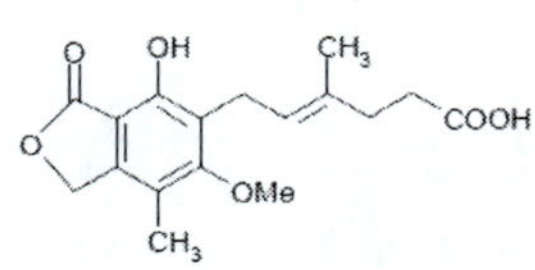

Mycophenolic acid

Figure 5. Structures of substances of polyketide nature.

Benzodiazepine alkaloids cyclopeptine and cyclopenine were also identified (Figure 3). Quinoline alkaloids were represented by viridicatine and quinocitrinines; quinazoline alkaloids were represented by fumiquinozalines F and G (Figure 4). Substances of polyketide nature were identified: ochratoxins A and B, mycophenolic acid, griseofulvin, and PC-2 (Figure 5).

Thus, fungal strains that have existed under many years of cryoconservation conditions have been shown to produce secondary metabolites of various structural groups. By their biological activities, these substances are, in most cases, assigned to mycotoxins. Considering global warming forecasts, our data on the toxigenic potential of the fungi will be of interest to specialists dealing with human and animal health.

Secondary Metabolites from Permafrost and Their Use for Species Identification

The exact determination of permafrost fungus species at the moment of their isolation from natural substrates was found to be difficult due to the shift of the optimum temperature of growth on diagnostic agar media to lower temperatures (Zhelifonova et al., 2009; Kozlovsky et al., 2003a). The results of work on successive inoculations of cultures led us to conclude that the growth rates of most studied strains increased, which motivated the redetermination of the species of several strains of the subgenera *Penicillium* and *Furcatum*. The species attribution for the *Aspergilloides* and *Biverticillium* subgenera remained as it was initially (Zhelifonova et al., 2010).

Recently, a new system was developed for the identification of fungi of the *Penicillium* subgenus which, along with the micro- and macromorphological characters of the studied strains, makes use of their secondary metabolite profiles (Samson and Frisvad, 2004). This system was used to refine the taxonomic positions of *Penicillium* strains isolated from usual habitats (Samson and Frisvad, 2004; Houbraken et al., 2011).

The use of secondary metabolites in taxonomy is based on empirical observations of general physiological and biochemical characteristics in phylogenetically related organisms. Potential and factual production of secondary metabolites is part of physiologico-biochemical identification. Proceeding from this approach and based on secondary metabolite profiles, the species names were refined for 41 *Penicillium* strains (Antipova et al., 2011; Kozlovsky et al., 2012).

Most *Penicillium* subgenus strains were attributed to the section *Viridicata*. The identification of species in this section is often problematic due to the existence of common interspecies morphological characters, in the context of which it makes special sense to determine the ability of strains to produce secondary metabolites of diagnostic significance.

Correspondence was observed between species names and sets of morphological characters and secondary metabolites known for the species *P. verrucosum* Dierckx and *P. griseofulvum* Dierckx. Thus, the occurrence of ochratoxins A and B in five strains unambiguously indicated their attribution to the species *P. verrucosum* (Zhelifonova et al., 2009). The griseofulvin identified in one of these strains corresponded to the initially determined species *P. griseofulvum* (Kozlovsky et al., 2012). For the other 41 strains, the characters were noted to be not in correspondence. By their secondary metabolite profiles, the cultures were divided into several groups. Each group of strains adapted to artificial storage conditions was characterised by an identical set of macro- and micromorphological characters. The group synthesising only CPA included 10 strains (Table 1). Their morphological characters corresponded to fungi of the section *Viridicata*, series *Camemberti*. An ability to produce only CPA was characteristic of only one strain, *P. camemberti* Thom. In other CPA-producing *Penicillium* species of the series *Camemberti*, other compounds acted as chemotaxonomy markers: for *P. commune* Thom, these were rugulovasines; for *P. palitans* Westling, fumigaclavines (Kozlovsky et al., 2012). The morphological characters exhibited by these strains on diagnostic agar media were similar and corresponded to the species *P. commune.* The full set of secondary metabolites was identified in two strains characteristic of the species *P. palitans*: CPA, fumigaclavines A and B, and festuclavine. In five strains, fumigaclavines A and B and festuclavines were identified, but CPA was absent (Kozlovsky et al., 2009). The morphological characters exhibited by these strains on diagnostic agar media were very similar and corresponded to the species *P. palitans*.

Diketopiperazine alkaloids of the roquefortine group were identified in eight strains first attributed to *P. aurantiogriseum* Dierckx and *P. miczynskii* Zaleski (Antipova et al., 2011; Kozlovsky et al., 2003a). The species *P. aurantiogriseum* was attributed to the subgenus *Penicillium* section *Viridicata* series *Viridicata.* In this taxon, the biosynthesis of roquefortine-group metabolites was characteristic of only the species *P. melanoconidium* Frisvad & Samson. For *P. miczynskii* (subgenus *Furcatum*), the biosynthesis of roquefortine-group metabolites is unknown. Repeated identification of these cultures on diagnostic media and biosynthesis of roquefortine-group metabolites by them unambiguously indicated their attribution to the subgenus *Penicillium* section *Chrysogena* series *Chrysogena* species *P. chysogenum* Thom.

Producers of benzodiazepine (cyclopenine, cyclopeptine, and cyclopenol) and quinoline (viridicatine) alkaloids were initially assigned to the species of the series *Viridicata* (Kozlovsky et al., 2012). Cyclopenines and viridicatines served as chemotaxonomic markers of the species *P. cyclopium* Westling, *P. freii* Frisvad & Samson, *P. neoechinulatum* (Frisvad, Filtenborg & Wicklow) Frisvad & Samson, and *P. polonicum* Zalesky; and in the series *Solita*, of the species *P. discolor* Frisvad & Samson, *P. echinulatum* Fassatiova, and *P. solitum* Westling (Samson and Frisvad, 2004). The morphological characters exhibited on diagnostic media by the cultures adapted to artificial storage conditions and the biosynthesis of cyclopenines and viridicatines enabled these strains to be assigned to the species *P. solitum*. For other species producing cyclopenines and viridicatines, marker metabolites included other compounds.

Mycophenolic acid was identified in two strains attributed by their micro- and macromorphological characters to *P. brevicompactum* Dierckx (subgenus *Penicillium* section *Corotata* series *Olsoni*) (Zhelifonova et al., 2010). However, the production of only mycophenolic acid by these strains indicated their attribution to another species of the series *Olsoni*, *P. bialowiezense* Zaleski. These species are difficult to distinguish by their morphological characters. The difference between them is in the production of breviaramide A by *P. brevicompactum* (Samson and Frisvad, 2004).

Two strains synthesised quinozaline alkaloids fumiquinazolines F and G as well as the polyketide PC-2 (Kozlovsky et al., 2012). On diagnostic agar media, these cultures showed a not very high growth rate and psychrotolerance but a high halotolerance. According to earlier published data (Samson and Frisvad, 2004), such ecophysiological properties are peculiar to four species of *Penicillium* from the section *Viridicata* series *Verrucosa* (*P. verrucosum*, *P. thymicola*), *Camemberti* (*P. camemberti*), and *Corymbifera* (*P. radicicola* Overy & Frisvad).

The revealed production of fumiquinozalines and the micromorphological characters confirmed their attribution to the species *P. thymicola*. *P. thymicola* is a new species attributed by Frisvad and Samson (2004) to section *Viridicata* series *Verrucosa*. From the closely related species *P. verrucosum* and *P. nordicum* Dragoni and Marino, this species differed in the composition of metabolites produced, the rough conidia, and yellow reverse on Czapek Yeast Autolysate (CYA) agar.

Five out of 11 investigated strains of the species *P. variabile* Sopp (subgenus *Biverticillium*) isolated from various permafrost regions were found to produce clavine-group ergot alkaloids rugulovasines A and B (Zhelifonova et al., 2006; Antipova et al., 2008). The production of rugulovasines by fungi of this species was detected for the first time. Rugulovasines are synthesised by *Penicillium* fungi of two subgenera – *Biverticillium* (*P. purpurogenum* Stoll, *P. rugulosum* Thom, and *P. islandicum* Sopp) and *Penicillium* (*P. atramentosum* Thom, *P. biforme* Thom [= *P. camemberti*], *P. caseifulvum* Lund, and *P. commune* Thom; Samson and Frisvad, 2004). Close macro- and micromorphological characters in the species *P. variabile* and *P. rugulosum* can be a basis for species redetermination of the strains that synthesise rugulovasines. However, the attribution of the investigated rugulovasine producers to *P. variabile* raises no doubts due to the absence of conidial germination at 5°C.

We can assign to this group two strains that synthesised an identical range of metabolites consisting of ergolenes with a double bond at position 8,9: agroclavine-1, epoxyagroclavine-1, and quinoline alkaloids quinocitrinines A and B (Table 1). By their morphological characters, the strains were attributed to the species *P. citrinum* Thom and *P. waksmanii* Zaleski (Kozlovsky et al., 2003b; Kozlovsky et al., 2003c; Antipova et al., 2011). Similar metabolites were found in two cultures, *P. corylophillum* Dierckx and *P. fellutanum* Biourge, isolated from modern habitats (Kozlovsky et al., 2011). All these *Penicillium* species are attributed to the subgenus *Furcatum.*

Correlation of Secondary Metabolite Biosynthesis with Producer Growth

One of the important issues in the formation of secondary metabolites is the correlation between their biosynthesis and producer growth. For many producers of mainly polyketide-nature metabolites, penicillin, etc., the biosynthesis of secondary metabolites was shown to occur only in the exponential growth phase (Betina et al., 1995). In alkaloid producers, biosynthesis starts from the first day of growth and proceeds in parallel with the growth of the producer (Reshetilova and Kozlovsky, 1990). A similar tendency is preserved for fungi isolated from permafrost. Thus, in *P. thymicola*, yellow-brown pigments of polyketide nature began to be synthesised in the phase of active growth from day 7 and reached maximum values in the stationary phase, which is consistent with the classical idiophase model known for various secondary metabolites, but the biosynthesis of fumiquinazolines proceeded in parallel with fungal growth (Zhelifonova et al., 2012). Producers of alkaloids feature a cyclic accumulation of metabolites in the medium and in the mycelium, which is connected to culture growth phases. In *P. citrinum* VKM FW-800 isolated from permafrost, a decrease in the concentrations and in the specific rate of the accumulation of agroclavine-1, epoxyagroclavine-1, and quinocitrinines A and B in the culture liquid was observed in the second lag phase of fungal growth due to diauxia, and in the onset of the stationary phase (Kozlovsky et al., 2005). Model experiments have shown that the drop in the concentration of alkaloids in the culture liquid at the onset of the stationary growth phase is not connected to the transformation of these metabolites to others outside the cells but is caused by their absorption by the fungal cells (Kozlovsky and Zhelifonova, 2007). Dominant absorption of alkaloids from the culture liquid with respect to their excretion is possibly related to the interruption of alkaloid biosynthesis due to the drop in the level of free tryptophan in *P. citrinum* mycelium in the stationary growth phase (Kozlovsky et al., 2006). To check this assumption, tryptophan was added at concentrations of 0.25 mM and 2 mM during cultivation on day 10 or 11 before and during the drop in alkaloid concentration. Within 48 hours of observation, there was no biomass increase. The uptake of amino acid from the medium and its accumulation in the mycelium were detected. The addition of tryptophan led to enhancement of the biosynthesis of alkaloids by cells and inhibited their excretion into the cultivation medium. In studies of ergot alkaloid biosynthesis by *Claviceps* fungi, these compounds were assumed to bind to excess tryptophan, the precursor of alkaloids, thus preventing repression by the end product and the cessation of growth. In the case of *P. citrinum*, an inverse process possibly takes place – alkaloids are consumed at the onset of the stationary growth phase due to the low level of tryptophan required for metabolism. (Rehachek and Sajdl, 1990)

COMPARISON OF SECONDARY METABOLITE PRODUCERS ISOLATED FROM VARIOUS HABITATS

The biosynthetic capabilities of two strains isolated from permafrost, *P. citrinum* and *P. waksmanii,* and two strains isolated from modern habitats, *P.*

corylophilum and *P. fellutanum,* all of which are producers of epoxyagroclavine-1 and quinocitrinines, were compared (Kozlovsky et al., 2011). In *P. corylophilum* and *P. fellutanum*, the accumulation of quinocitrinines was found to be insignificant and not to exceed 1 mg/l, which was 10–15 times lower than that for the permafrost strains. It should be noted that both branches of the biosynthesis of ergoline- and quinoline-nature compounds are biogenetically connected by the shikimate pathway that forms their precursors tryptophan and anthranilic acid.

The addition of zinc ions to the cultivation media of the four producers was found to have different effects on epoxyagroclavine-1 and quinocitrinine biosynthesis. A stimulating effect of zinc ions on the biosynthesis of epoxyagroclavine-1 and quinocitrinine was observed only in *P. citrinum*. In the other producers, only the content of epoxyagroclavine-1 increased; the accumulation of quinocitrinine remained at approximately the same level, which indicates that epoxyagroclavine-1 and quinocitrinine biosynthesis genes are in different clusters.

Thus, the strains under permafrost conditions are more efficient producers of quinocitrinine antibiotics compared with strains with a different way of development. The most active strain *P. citrinum* is the most ancient (1.8–3 million years old); it is probable that the many years of cryoconservation conditions has led to the preservation of the strains' biosynthetic capabilities. Confirmation this are the data obtained in studies of passages and storage of *P. citrinum* culture. At successive inoculations of the strain to agar media, the culture was observed to dissociate; secondary growth colonies emerged and alkaloid production decreased. Storage of the culture on wort agar slants also resulted in a significant decrease in epoxyagroclavine-1 and quinocitrinine biosynthesis. The most preferable storage technique for the culture is lyophilisation.

The biosynthesis of clavine alkaloids rugulovasine A and B was studied in 18 *P. variabile* strains isolated from permafrost regions in Arctic and Antarctica and from modern habitats (Antipova et al., 2008; Zhelifonova et al., 2006). It was shown that rugulovasines were synthesised by 70% of strains isolated from modern habitats and 45% of strains isolated from permafrost. It should be pointed out that the level of rugulovasine accumulation in the strains isolated from modern habitats was 3–7 times higher than in the most active from the ancient strain *P. variabile* BKM FW-806. Differentiation of the mycelium and microcyclic conidiogenesis occurred in the early stages of growth of all rugulovasine producers. Thus, it was shown that the ability of rugulovasine biosynthesis in *P. variabile* arose before reaching the conditions of long-term cryopreservation.

Conclusion

Strains of *Penicillium* fungi isolated from permafrost remain highly capable of synthesising secondary metabolites of various structural groups. The cultures are producers of clavine, diketopiperazine, benzodiazepine, quinoline, quinozaline alkaloids, and substances of polyketide nature. By their biological activities, these substances are, in most cases, assigned to mycotoxins, for example, CPA, roquefortine, ochratoxins A and B, and mycophenolic acid. Considering global warming forecasts, our data on the toxigenic potential of the fungi will be of interest to specialists dealing with human and animal health.

A characteristic feature of the producer strains is the cyclicity of biosynthesis and the consumption of alkaloids, which is obviously related to maintaining intracellular tryptophan at a required level. Enzymes of secondary metabolite biosynthesis are of a constitutive nature. The secondary metabolite profiles are successfully used for purposes of taxonomy.

There is no significant difference between the profiles of the secondary metabolites of *Penicillium* fungi isolated from permafrost-bearing and permafrost-free habitats. It is known that the genes for the biosynthesis of secondary metabolites are organised in clusters found in chromosome DNA. It is probable that during this period of time, no evolutionary changes in the genome controlled the secondary metabolite biosynthesis taken place in the strains the past another way of the development. It seems, during this period of time there have been no evolutionary changes in the genome, responsible for the biosynthesis of secondary metabolites of fungi strains, another way past development.

References

Antipova, T.V., Zhelifonova, V.P., Kochkina, G.A., Kozlovskii, A.G. (2008) Growth and biosynthesis of rugulovasines in *Penicillium varicbile* Sopp 1912. *Mikrobiol.* 77, 502 – 507.

Antipova, T.V., Zhelifonova, V.P., Baskunov, B.P., Ozerskaya, S.M., Ivanushkina, N.E., Kozlovsky A.G. (2011) New producers of biologically active compounds—fungal strains of the genus *Penicillium* isolated from permafrost. *Prikl. Biokhim. Mikrobiol.* 47, 288 – 292.

Betina, V. (1995) Differentiation and secondary metabolism in some prokaryotes and fungi. *Folia Microbiol.* 40, 51 – 67.

Dreyfuss, M.M., Chapela, I.H. (1994) Potential of fungi in the discovery of novel, low-molecular weight pharmaceuticals. *The Discovery of Natural Products with Therapeutic Potential.* Boston-London: Buttererworth-Heinemann Press, 49 – 79.

Gilichinsky, D. Rivkina, E., Bakermans, C., Shcherbakova, V., Petrovskaya, L., Ozerskyaya, S., Ivanushkina, N., Kochkina, G., Laurinavichius, K., Pecheritsina, S., Fattakhova, R., Tiedje, J.M. (2005) Biodiversity of cryopegs in permafrost. *FEMS Microbiol. Ecol.* 53, 117 – 128.

Gilichinsky, D.A., Wilson, G.S., Friedmann, E.I., McKay, C.P., Sletten, R.S., Rivkina, E.M., Vishnivetskaya, T.A., Erokhina, L.G., Ivanushkina, N.E., Kochkina, G.A., Shcherbakova, V.A., Soina, V.S., Spirina, E.V., Vorobyova, E.A., Fyodorov-Davydov, D.G., Hallet, B., Ozerskaya, S.M., Sorokovikov, V.A., Laurinavichyus, K.S., Shatilovich, A.V., Chanton, J.P., Ostroumov, V.E., Tiedje, J.M. (2007) Microbial populations in Antarctic permafrost: biodiversity, state, age, and implication for astrobiology. *Astrobiol.* 7, 275 – 311.

Houbraken, J., Frisvad, J. C., Samson, R.A. (2011) Taxonomy of *Penicillium* section *Citrina*. *Stud. Mycol.* 70, 53 – 138.

Kochkina, G.A., Ivanushkina, N.E., Karasev, S.G., Gavrish, E.Yu., Gurina, L.V., Evtushenko, L.I., Spirina, E.V., Vorob′eva, E.A., Gilichinskii, D.A., Ozerskaya, S.M. (2001) Survival of Micromycetes and Actinobacteria under condition of long-term natural cryopreservation. *Mikrobiol.* 70, 412 – 420.

Kozlovsky, A.G., Zhelifonova, V.P., Adanin, V.M., Antipova, T.V., Ozerskaya, S.M., Ivanushkina N.E., Graefe U. (2003a). *Penicillium aurantiogriseum* Dierckx 1901: Producer of diketopiperazine alkaloids (roquefortine and 3,12-dihydroroquefortine), isolated from permafrost. *Prikl. Biokhim. Mikrobiol.* 39, 393 – 397.

Kozlovsky, A.G., Zhelifonova, V.P., Antipova, T.V., Adanin, V.M., Ozerskaya, S.M., Kochkina, G.A. (2003b) Quinocitrinines A and B, new quinoline alkaloids from *Penicillium citrinum* VKM FW-800, a permafrost fungus. *J. Antibiot.* 56, 488 – 491.

Kozlovsky, A.G., Zhelifonova, V.P., Adanin, V.M., Antipova, T.V., Ozerskaya, S.M., Kochkina, G. A., Graefe, U. (2003c) The Fungus *Penicillium citrinum* Thom 1910 VKM FW-800 isolated from ancient permafrost sediments as a producer of the ergot alkaloids agroclavine-I and epoxyagroclavine-I. *Mikrobiol.* 72, 723 – 727.

Kozlovsky, A.G., Zhelifonova, V.P., Antipova, T.V. (2005) The fungus *Penicillium citrinum*, isolated from permafrost sediments, as a producer of ergot alkaloids and new quinoline alkaloids quinocitrinines. *Prikl. Biokhim. Mikrobiol.* 41, 568 – 572.

Kozlovsky, A.G., Zhelifonova, V.P., Antipova, T.V. (2006) The effect of some amino acids on the growth and biosynthesis of ergot alkaloids and quinocitrinines in the fungus *Penicillium citrinum*. *Mikrobiol.* 75, 334 – 341.

Kozlovsky, A.G., Zhelifonova, V.P. (2007) Transport of ergot alkaloids and quinocitrinins in the producing fungus Penicillium citrinum *Mikrobiol.* 76, 456– 471.

Kozlovsky, A.G., Zhelifonova, V. P., Adanin, V.M., Antipova, T.V. (2009) Clavine alkaloid biosynthesis by the fungus *Penicillium palitans* Westling 1911 isolated from ancient permafrost deposits. *Prikl. Biokhim. Mikrobiol.* 45, 202 – 206.

Zhelifonova, V. P., Antipova, T.V., Kozlovsky, A.G. (2010) Secondary metabolites in taxonomy of the *Penicillium* fungi. Mikrobiol. 79, 277-286.

Kozlovsky, A.G., Zhelifonova, V. P., Antipova, T.V., Zelenkova N.F. (2011) Physiological and biochemical characteristics of fungi of the genus *Penicillium* as producers of ergot alkaloids and quinocitrinines. *Prikl. Biokhim. Mikrobiol.* 47, 469 – 473.

Kozlovsky, A.G., Zhelifonova, V.P., Antipova, T.V., Baskunov, B.P., Kochkina, G.A., Ozerskaya, S.M. (2012) Secondary metabolite profiles of the *Penicillium* fungi isolated from the Arctic and Antarctic permafrost as elements of polyphase taxonomy. *Mikrobiol.* 81, 332 – 338.

Ozerskaya, S., Kochkina, G., Ivanushkina, N., Gilichinsky, D. (2004) Mycelial fungi in cryopegs. *Int. J. Astrobiol.* 3, 327 – 331.

Ozerskaya, S., Kochkina, G., Ivanushkina, N. (2009) Fungi in permafrost. In: R. Margesin (ed.) Permafrost Soils. Soil Biology. Springer-Verlag Berlin Heidelberg; 16, 85 – 95.

Pitt, J.I. (1979) The genus *Penicillium* and its teleomorphic states *Eupenicillium* and *Talaromyces*. Acad. Press, London.

Rehachek Z., Sajdl P. (1990) Ergot Alkaloids: Chemistry, Biological Effect, Biotechnology / Ed. J. Spoutsova. Academia, Praha.

Reshetilova, T.A., Kozlovsky, A.G. (1990) Biosynthesis alkaloids by filamentous fungi. *Prikl. Biokhim. Mikrobiol.* 26, 291 – 306.

Robinson, C.H. (2001) Cold adaptation in Arctic and Antarctic fungi. *New Phytol.* 151, 341 – 353.

Samson, R.A., Frisvad, J.C. (2004) *Penicillium* subgenus *Penicillium*: New taxonomic schemes, mycotoxins and other extrolites. *Stud. Mycol.* 49, 1 – 174.

Wagner, D. (2008) Microbial communities and processes in Arctic permafrost environments. In: P. Dion, C.S. Nautiyal (eds) Microbiology of extreme soils. Soil Biology Springer-Verlag Berlin Heidelberg; 13, 133 –154.

Willerslev, E., Hansen, A.J., Poinar, H.N. (2004) Isolation of nucleic acids and cultures from fossil ice and permafrost. *Trends Ecol. Evol.* 19, 141 – 147.

Zhelifonova, V.P., Antipova, T.V., Kozlovsky, A.G. (2006) The fungus *Penicillium variabile* Sopp1912 isolated from permafrost deposits as a produce of rugulovasines. *Mikrobiol.* 75, 742 – 746.

Zhelifonova, V.P., Antipova, T.V., Ozerskaya, S.M., Kochkina, G.A., Kozlovsky, A.G. (2009) Secondary metabolites of *Penicillium* fungi isolated from permafrost deposits as chemotaxonomic markers. *Mikrobiol.* 78, 350 – 354.

Zhelifonova, V.P., Antipova, T.V., Kozlovsky, A.G. (2010) Secondary metabolites in taxonomy of the *Penicillium* fungi. *Mikrobiol.* 79, 277 – 286.

Zhelifonova, V. P., Antipova, T.V., Kozlovsky, A.G. (2012) Biosynthesis of fumiquinazolines by the fungus *Penicillium thymicola. Prikl. Biokhim. Mikrobiol.* 48, 334 – 339.

Zucconi, L., Selbamann, L., Buzzini, P., Turchetti, B., Guglielmin, M., Frisvad, J.C., Onofri S. (2012) Searching for eukaryotic life preserved in Antarctic permafrost. *Polar. Biol.* 35, 749 – 757.

INDEX

#

A

B

C

D

G

H

I

J

K

L

M

N

O

P

R

T

U

V

W

Y

Z